KV-506-774

Microbiologically influenced corrosion handbook

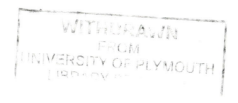
WITHDRAWN
FROM
UNIVERSITY OF PLYMOUTH
LIBRARY

Microbiologically influenced corrosion handbook

SUSAN WATKINS BORENSTEIN

WOODHEAD PUBLISHING LIMITED

Cambridge, England

UNIVERSITY OF PLYMOUTH

Item No. 900 275657-9

Date 1 9 JUL 1996 (T)

Class No. 620.11223 BOR

Contl. No. 1855731274

LIBRARY SERVICES

Published by Woodhead Publishing Ltd,
Abington Hall, Abington,
Cambridge CB1 6AH, England

First published 1994, Woodhead Publishing Ltd

© Woodhead Publishing Ltd

Conditions of sale
All rights reserved. No part of this publication may be
reproduced or transmitted in any form or by any means,
electronic or mechanical, including photocopy, recording, or any
information storage and retrieval system, without permission in
writing from the publisher.

British Library Cataloguing in Publication Data
A catalogue record for this book is available from the British Library

ISBN 1 85573 127 4

Designed by Geoff Green (text) and Chris Feely (jacket).
Typeset by Best-set Typesetter Ltd., Hong Kong.
Printed in the United States of America.

Contents

Preface

In 1986, Fontana estimated that corrosion costs industry a staggering \$126 billion per year.[1] Most metal alloys corrode in water, as anyone who has ever seen a ship's hull or marine hardware close to can tell you. The culprits are neither wholly salt nor barnacles, however, but, to varying degrees, microorganisms, particularly certain bacteria and fungi. Such corrosion is termed microbiological influenced corrosion, or MIC.

When a metallic surface is immersed in water, even potable water, microorganisms immediately attach themselves to that surface, develop a biofilm – a microbial mass composed of aquatic bacteria, algae and other microorganisms – and proceed to produce by-products, as part of their natural metabolic processes. The result for the microorganism is sustained life; for the metal surface, corrosion.

In recent years, analyses of costly pitting failures at nuclear power plants, chemical process plants, and pulp and paper mills have shown that stainless steel is particularly susceptible to MIC, in the form of pitting to welds.[2] While pitting by its nature is extremely localized, pitting failures can be devastating, and require costly repairs.

This text provides fundamental background for understanding the interdisciplinary roles of microbiology, metallurgy and electrochemistry as they relate to MIC. Methods by which MIC can be detected and monitored are discussed, as well as its prevention. We will also consider how welding, heat treatment and other metallurgical and process variables affect corrosion resistance.

There are many misunderstandings about MIC; there are even those who do not believe MIC actually occurs. The time is ripe for a thorough investigation of this little-understood, yet pervasive and costly, form of corrosion. For anyone interested in understanding and mitigating the effects of MIC, whether in industry or research, I hope this handbook will be instructive.

References

1 Fontana, M., 1986, *Corrosion Engineering*, McGraw-Hill, New York.
2 Institute of Nuclear Power Operations, 1984, 'Microbiologically Influenced Corrosion (MIC)', Institute of Nuclear Power Operations Significant Event Report, SER 73-84.

Acknowledgements

I am deeply indebted to the many people who reviewed this manuscript for technical accuracy:

- Naval Research Laboratory, Stennis Space Center, MS: Dr Brenda Little (SEM and ESEM micrographs); Patricia Wagner (SEM and ESEM micrographs); Richard Ray (SEM and ESEM micrographs).
- University of Southern California, Los Angeles, CA: Dr Florian Mansfeld.
- Structural Integrity, San Jose, CA: George Licina.
- Corrosion Engineering and Research Company, Concord, CA: J. Darby Howard; John French; Bruce Kelley.
- Cortest Columbus Testing, Columbus, OH: Dr John Beavers (electrochemistry); Dr Neil Thompson (electrochemistry).
- Puckorius and Associates, Evergreen, CO: Paul Puckorius (water treatment).
- Bechtel Group, Inc., San Francisco, CA: Dr Richard A. White; Yun Chung (photomicrographs of stainless steel pits); Curt Cannell (photomicrographs of stainless steel pits).
- Aptech Engineering, Sunnyvale, CA: Philip Lindsay.
- Montana State University, Bozeman, MT: Dr Gil Geesey (microbiology and photomicrographs).
- Naval Warfare Center, Silver Springs, MD: Dr Joanne Jones-Meehan (microbiology and photomicrographs); Dr Maryanne Walch (microbiology and photomicrographs).
- Harvard University, Cambridge, MA: Dr Timothy Ford (microbiology).
- Colorado School of Mines, Golden, CO: Dr J. Jones (welding and metallurgy).
- Metallurgical and Welding Consulting Services, Sunnyvale, CA: Donna Schubert.
- Allegheny Ludlum Steel Corp., Brackenridge, PA: Jeffery Kearns.
- ANR Pipeline, Dearborn MI: Rick Eckert.
- MIC Associates, Chadds Ford, PA: Robert Tatnall.

- Bioindustrial Technology, Inc., Georgetown, TX: Dr Daniel Pope (case histories).

My warmest thanks are also due Marina Hirsch, editor and friend, and Clare Markovits and Jan Ferris, technical reviewers and text processors.

The never-ending encouragement of Daniel Borenstein, my wonderful, supportive husband, and Crystal Knight, my loving daughter, made this book possible.

Chapter 1

Introduction to microbiologically influenced corrosion

This chapter describes the relationship of destruction and degradation primarily of metal surfaces by various life forms.

1.1 Background

When a metal is exposed to natural waters, corrosion begins immediately. In some cases, microorganisms influence the corrosion process. The corrosion of the metal and the composition of the corrosion deposits largely depend on:[1]

- Microbiology.
- Metallurgy.
- Electrochemistry.
- Bulk water chemistry.

Microbiology is a factor in terms of environmental variables and the activity of the organisms, metallurgy is a key factor, given the many variables, such as the type, structure and processing of the metal and electrochemistry is essential because of the mechanisms by which the corrosion processes occur. For example, the metal goes into solution as ions, and solid products, such as rust, form by a subsequent reaction.

Not all corrosion is MIC. Resolving, understanding and mitigating MIC problems require familiarity with all four fields: microbiology, metallurgy, electrochemistry and water chemistry.

1.2 Biofilms

The instant a metallic surface is immersed in water, a biofilm begins to form.[2] A biofilm is a microbial mass composed of aquatic bacteria, algae and other microorganisms.

1.2.1 Development of a biofilm

Many species of bacteria, algae and fungi bind various metals and form biofilm.[3] The development of that film occurs in four stages:[4]

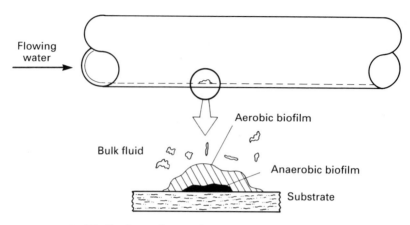

1.1 Biofilm formation on the inside surface of a pipe.

- Conditioning: instantaneous chemical adsorption of organics; organisms not directly involved.
- Adhesion by 'pioneer' bacterial species: bacterial epiphytes adhere to submerged surfaces in a matter of hours.
- Colonization of other microorganisms: other bacteria and fungi become associated with the surface following colonization by the pioneering species over a matter of days.
- Accumulation: entrapment of particles, dead cells and chelation of heavy metals from water (both as corrosion products and ions in bulk solution).

Figure 1.1 is a sketch of a biofilm formed on a metal surface. A biofilm begins with the adsorption of organic matter on the metal surface from the bulk environment[5] (to adsorb means to collect, e.g. a gas, liquid or dissolved salt, in a condensed form on a surface). The turbulent flow transports microbes to the surface, and the microorganisms attach and then grow, using nutrients from the water. The film may eventually be sheared away with the flowing water. These steps are shown in Fig. 1.2.[5] Geesey's work shows the development of biofilms on several metal surfaces in Fig. 1.3.[3]

Microbiological cells are usually either sessile (attached within a biofilm, usually to something solid and immobile) or planktonic (move with flow of water). These terms describe a generalized water system, in which some microorganisms are attached to the walls, and some float freely in the bulk environment.

Microbiological cells are usually found in three states: planktonic; sessile; and a subset of the others termed fragments, often called sessile particulates.[6] Planktonic cells are dispersed in the aqueous phase, while sessile cells are immobilized with respect to the surrounding environment.

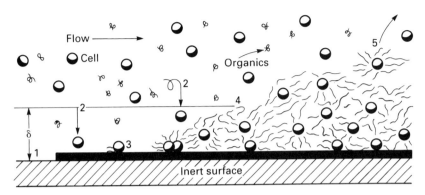

1.2 Steps in biofilm formation. Formation is initiated when small organic molecules become attached to an inert surface (1) and microbiological cells are adsorbed onto the resulting layer (2). The cells send out hairlike exopolymers to feed on organic matter (3), adding to the coating (4). Flowing water detaches some of the formation (5), producing an equilibrium layer δ (source: ASM Metals Handbook, 1987, Vol. 13, *Corrosion*, ASM International, Metals, Park, OH, p. 492).

In soils and water, most microorganisms are found in the sessile rather than the planktonic state. Sessile microorganisms usually form a biofilm through the elaboration of extracellular polymer.[7] The polymers bond the cells together and protect them from any hostile conditions that may arise in the environment. Biofilms retain water, accumulate nutrients and form a polymeric matrix. Extracellular polymers are discussed in Section 2.4.1.

1.3   Biofouling

Biofouling is a general term describing all forms of biological growth on surfaces in contact with natural waters.[8] Characklis defines fouling as the undesirable formation of deposits on equipment surfaces, which significantly decreases equipment performance and/or the useful life of the equipment.[9] Several types may occur: biological; corrosion; particulate; precipitation; or combinations thereof.

Biofouling can reduce the flow of water through heat exchangers, which in turn reduces the heat transfer properties and may degrade entire systems, by loss of efficiency.[5]

Cooling water systems are attractive locations for microbial growth and biofilm formation. Such systems can act as incubators. Consider the conditions: the water is highly oxygenated, exposed to sunlight and air, and maintained at a temperature of 27–60 °C (80–140 °F), and a pH between 6 and 9. These conditions promote biofouling of system surfaces in contact with the water.[5]

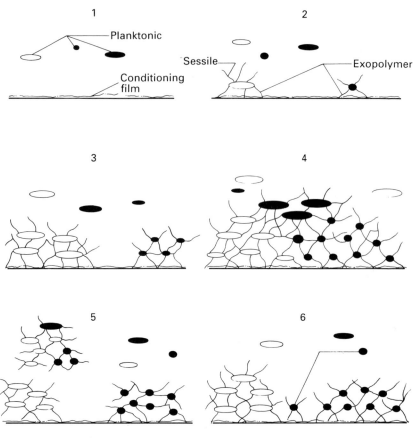

1.3 Biofilm development.

1.4   Biocorrosion

When biofouling leads to the corrosion of metal surfaces, that process in often termed biocorrosion. The terms biocorrosion and MIC are often used interchangeably. Figure 1.4 is a representation of what occurs when active corrosion-causing bacteria develop on a metal surface.[10]

- A thick biofilm develops. An anaerobic zone develops adjacent to the colonized surface.
- Microorganisms develop colonies and complex consortia. These trap ions and create localized chemical and physical gradients at the metal surface.
- An electrochemical cell is set up and the metal dissolves, causing a pit to occur beneath the area affected.

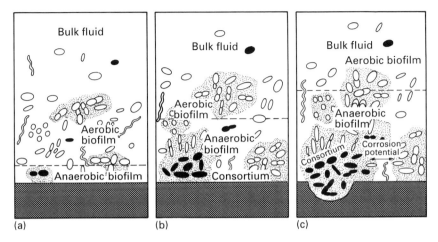

1.4 Development of a biofilm: (a) a variety of microorganisms colonize a surface; (b) microorganisms develop colonies and consortia; (c) an electrochemical cell is set up causing pitting to occur (source: Costerton, J.W., Geesey, G.G., and Jones, P.A., 1987, 'Bacterial biofilms in relation to internal corrosion monitoring and biocides', *Corrosion/87*, paper no. 57, NACE, Houston, TX, reproduced with permission).

There are three very important aspects:[11]

- No new type of corrosion reaction is caused by the presence of the microbes.
- Corrosion is strongly associated with the biofilm and extracellular polymer.
- Most engineering alloys are susceptible to this problem.

1.5   Biodeterioration

As defined by Hueck, biodeterioration is the biological mediated break-down or destruction of materials of economic importance by organisms. It should not be confused with biodegradation, which is the breaking down of materials by organisms resulting in improved quality.[12, 13] For example, detoxifying waste pesticides is commonly described as biodegrading them. The difference between the two processes is the intent. When deterioration is unintended, it is biodeterioration; when it is intended, it is biodegradation.[13]

In keeping with his broad definition, as above, Hueck broadly classifies the biodeterioration process into three types:[12, 13]

- Mechanical: insect and rodent attack on non-nutrient materials such as pipe.

- Chemical: either similar, in which the material is a food source, or dissimilar, where the waste produced (such as metabolites) degrades the material but is not used as a food source by the organism.
- Fouling and soiling: where the organism causes a worsening of the material, for example, by its mere presence or by its secretions.

This classification is useful for determining control methods for the biodeterioration problem, but does not reflect on the metabolic activities of the organism.[12]

Suggested reading

1 Fleming, H.C., and Geesey, G.G., ed., 1991, *Biolouling and Biocorrosion in Industrial Water Systems*, Springer-Verlag, New York.
2 Characklis, W.G., Marshall, K.C., ed., 1990, *Biofilms*, John Wiley, New York.
3 Characklis, W.G., Little, B.J., and McCaughey, M.S., 1989, in *Microbial Corrosion: 1888 Workshop Proceedings*, G. Licina, ed., EPRI ER-6345, Electric Power Research Institute, Palo Alto, CA.
4 Pope, D.H., Duquette, D., Wayner, P.C., and Johannes, A.H., 1989, *Microbiologically Influenced Corrosion: A State-of-the-art Review*, MTI Publication No. 1, Materials Technology Institute of the Chemical Process Industries, St Louis, MO.
5 ASM Metals Handbook, 1987, Vol. 13, *Corrosion*, ASM International, Metals Park, OH.
6 Howsam, P., ed., 1990, *Microbiology in Civil Engineering*, Proceedings of the Federation of European Microbiological Societies, FEMS Symposium No. 59, E. & F.N. Spon, University Press, Cambridge, UK.
7 Allsopp, D., and Seal, K.J., 1986, *An Introduction to Biodeterioration*, Edward Arnold, London.

References

1 Parr, J.G., and Hanson, A., 1965, *An Introduction to Stainless Steel*, ASM International, Metals Park, OH, p. 19.
2 Uhlig, H.H., 1948, *Corrosion Handbook*, John Wiley & Sons, New York.
3 Fleming, H.C., and Geesey, G.G., eds., 1991, *Biofouling and Biocorrosion in Industrial Water Systems*, Springer-Verlag, New York, p. 162.
4 Corpe, W.A., 1977, 'Marine microfouling and OTEC heat exchangers', in *Proceedings of the Ocean Thermal Energy Conversion, Biofouling and Corrosion Symposium*, ed. Gray, R.H., p. 31.
5 ASM *Metals Handbook*, 1987, Vol. 13, *Corrosion*, ASM International, Metals Park, OH, p. 492.
6 Howsam, P., ed., 1990, *Microbiology in Civil Engineering*, Proceedings of the Federation of European Microbiological Societies, FEMS Symposium No. 59, E. & F.N. Spon, University Press, Cambridge, UK, p. 6.
7 Costerton, J.W., and Lappin-Scott, H.M., 1989, Behaviour of bacteria in

biofilms, *American Society of Microbiology News*, **55**(12), 650–654.

8   Little, B.J., 1985, 'Succession in microfouling', in *Proceedings of the Office of Naval Research Symposium on Marine Biodeterioration*, U.S. Naval Inst. Press, Bathesda, MD.

9   Characklis, W.G., 1984, 'Biofilm development: a process analysis', in *Microbial Adhesion and Aggregation*, Marshall, K.C., ed., Springer-Verlag, New York, p. 137.

10 Costerton, J.W., Geesey, G.G., Jones, P.A., 1987, 'Bacterial biofilms in relation to internal corrosion monitoring and biocides', *CORROSION/87*, paper no. 57, National Association of Corrosion Engineers, Houston, TX.

11 Tiller, A.K., 1991, *Microbiology in Civil Engineering*, E. & F.N. Spon, London, p. 24.

12 Hueck, H.J., 1965, 'The biodeterioration of materials as a part of hydro-biology', *Material and Organisms* **1** 5–34.

13 Seal, K.L., 1991, 'Biodeterioration of materials used in civil engineering', in *Microbiology in Civil Engineering*, Proceedings of the Federation of European Microbiological Societies, FEMS Symposium No. 59, E. & F.N. Spon, London.

Microbiology

This chapter describes the relationship of microorganisms to metal corrosion. The many species of microorganisms and their metabolic diversity enable them to form support systems for cross-feeding, thus enhancing their survival.[1] For instance, microorganisms tolerate high pressures and a wide range of temperatures (subfreezing to above the boiling point of water), pH values (0 to 11) and oxygen concentrations (0 to almost 100%).[1] In addition, they can go into spore form under dehydrating conditions and then germinate later when conditions are acceptable for growth.

MIC commonly results when water remains in stainless steel components or piping systems after hydrostatic testing. (Hydrostatic testing involves filling a system with water and checking for leaks and structural integrity under pressure.) Although stagnant water conditions *per se* are not likely to produce direct corrosive attack given the inherent corrosion resistance of the materials, such conditions are ideal for MIC.

Natural water, and even potable water, contain a variety of bacteria. When a metallic surface is immersed in natural water, two processes occur simultaneously: corrosion starts immediately, and a biofilm begins to form.[2,3] A biofilm is a microbial mass composed of aquatic bacteria, algae and other microorganisms.

The microorganisms' metabolic processes are sustained by chemical reactions energized by nutrients obtained from the surrounding environment. These processes can influence the corrosion behaviour of materials by introducing or enhancing heterogenity at the surface by:[2,4]

- Destroying the protective films on the metal surface.
- Producing a localized acid environment.
- Creating corrosive deposits.
- Altering anodic and cathodic reactions, depending on the environment and organisms involved.

2.1   Glossary

The following are frequently used microbiological terms relevant to MIC.

2.1.1   General terms

- **Microbiologically influenced corrosion** (MIC): corrosion initiated or accelerated by microorganisms. MIC is also known as biological corrosion, biologically influenced corrosion, biologically induced corrosion, microbial corrosion, microbiologically induced corrosion and biocorrosion.
- **Biofouling**: all forms of biological growth on surfaces in contact with natural waters.[5]
- **Microfouling**: deposits caused by the growth of microbes, such as bacteria, and the corrosion products on surfaces.[5]
- **Biomass**: The collection of organisms combined with the deposits formed by microbial growth.

2.1.2   Water-related terms

- **Natural water**: fresh, brackish or salt.
- **Fresh water**: contains less than 1000 ppm of the chloride ion.
- **Brackish water**: contains roughly 1000 to 10 000 ppm of chlorides.
- **Salt water**: contains from 2.5% to 3.5% sodium chloride.
- **Bulk water**: water that supplies chemicals and microbes to system surfaces.

2.1.3   Organism-related terms

- **Aerobic**: having air or uncombined oxygen.
- **Anaerobic**: without air or uncombined oxygen.
- **Planktonic**: microscopic animal and plant life whose movements are controlled by water movement.
- **Sessile**: attached, such as attached to a surface.
- **Motile**: movement independent of water.
- **Prokaryote**: an organism lacking a true nucleus in the cell and that reproduces by binary fission.
- **Eukarycote**: an organism with a cell with a true nucleus that divides by mitosis or meiosis.
- **Heterotroph**: an organism that requires organic material for its source of cell carbon and energy.
- **Autotroph**: an organism that requires only carbon dioxide or carbonates as its source of carbon for making cell materials.
- **Hydrogenase**: an enzyme that catalyses the oxidation of hydrogen, and is possibly involved with the cathodic depolarization by sulphate-reducing bacteria.
- **Heterotroph**: an organism that depends on organic substances for its source of carbon.

- **Halophile**: an organism that grows well in high sodium chloride concentrations.
- **Mesophile**: an organism that grows well at temperature ranges of 25–40 °C (75–100 °F).
- **Barophile**: an organism that grows well at high pressures.
- **Obligate**: an organism that is restricted to a particular mode of life, such as a strict or obligate aerobe, which must have O_2 in the former case, or uncombined oxygen in the latter, to survive.
- **Facultative**: an organism that is not restrictive and can live in more than one condition. An example would be a facultative anaerobe, which prefers anaerobic conditions, but can live in some other condition, such as aerobic, if necessary.
- **Spore-forming**: an organism that forms spores. This process isolates the organisms from the environment (usually heat and dryness) until the organism is ready to regrow when the environment changes to a more favourable condition. Examples include two genera, *Bacillus* and *Clostridium*.
- **Electron-donor**: an animate or inanimate source of electrons (for instance, either a metal or a microorganism) that donates an electron.
- **Electron-acceptor**: an animate or inanimate acceptor of electrons.
- **Thermophile**: an organism that grows at temperatures above 45 °C (110 °F) (heat-loving).
- **Psychrophile**: an organism that grows at temperatures below 20 °C (70 °F) (cold-loving).
- **Mesophile**: an organism intermediately attracted to heat, 20–24 °C (70–75 °F).
- **Algae**: a group of aquatic organisms that contain chlorophyll. They are usually able to survive periods of dryness and are often found in soils, in tree barks, on rocks, or in coastal or moist climates.
- **Fungus**: a group of plants that lacks chlorophyll and includes moulds, rusts, mildew, smuts and mushrooms.
- **Gram's method**: a method of staining bacteria for classification purposes.
- **Gram-negative**: does not retain a colour stain; refers to cell envelope architechture (see Gram's method).
- **Gram-positive**: does retain a colour stain (see Gram's method).

2.1.4   Film terms

- **Biofilm**: a consortium of sessile organisms enveloped in exopolymer.
- **Exopolymer/extracellular polymer**: secretions that form a matrix of fibres (for example, snail slime).
- **Tubercle**: a localized mound on a metal surface.

2.1.5   Treatment terms

- **Biocide**: a chemical or compound, such as gluteraldehyde, which kills or reduces the number of organisms.
- **Chelates**: chemical compounds in which the central atom (usually a metal ion) is attached to neighbouring atoms by at least two bonds.
- **Sterilize**: to kill or reduce the population of organisms; to cause irreversible inactivation of life. The term sterile is misleading, since it carries a variety of meanings. Except under exacting conditions, such as making water for injection in the medical field, researchers are logistically unable to create truly 'sterile' water. They should aim towards reducing organisms to a very low number.
- **Biostat**: a compound that inhibits the growth but not necessarily the viability of organisms. Such compounds may be proprietary products and are often found in anti-fouling paints.

2.2   General properties of microorganisms

Microorganisms are organisms that encompass a wide range of sizes and morphologies but are commonly less than $1\,\mu m$ (usually about $0.5\,\mu m$) in length.[6] The physiological diversity is great.[7] Even when many chemical parameters are changed (such as lowering the pH), a large population of microorganisms will remain in the water and form biofilms that are detrimental to man-made structures.

The book *Microbiology for Environmental Scientists and Engineers*, by A. Gaudy and E. Gaudy, gives a thorough introduction to microbiology, focusing on problems related to pollution control. *Aquatic Microbiology: An Ecological Approach*, edited by T.E. Ford, provides an introduction to aquatic microbiology. *Biofouling and Biocorrosion in Industrial Water Systems*, edited by H.C. Fleming and G.G. Geesey, covers biofouling, microbial growth and contamination in water systems and biocorrosion.

2.2.1   Classifications of organisms

Scientists classify plants and animals in the following groupings:

- Kingdom.
- Phylum (division).
- Class.
- Order.
- Family.
- Genus.
- Species.

It can be confusing to apply this system to microorganisms. Some are not grouped into families, but by other methods, although in general they are classified by family, genus and species. Usually, bacteria are grouped by identical morphologies and behaviour.

A genus is a group of related species. Species with many characteristics in common belong to the same genus. A family is a group of related genera. Bacteria have two names: genus is their first, species their second. For example, *Escherichia coli* (genus, species) is commonly known as *E. coli* and is possibly the most thoroughly studied microorganism. It is an enterobacteria, and can cause infections in humans, as well as live in the intestines of humans. Classification tells the characteristics of that type of organism.

Microorganisms fall into several groupings that include:

● Protozoa (of the kingdom Protista).
● Algae.
● Fungi.
● Bacteria.
● Viruses.

A new taxonomy based on t-RNA base sequences has changed bacterial identification. Microorganisms fall into several groupings that include:

● Protozoa.
● Algae.
● Fungi.
● Bacteria.
● Viruses.

There are two classifications for organisms: prokaryote and eukarycote. A prokaryote is an organism lacking a true nucleus to the cell and reproducing by fission. A eukarycote is an organism made up of cells with a true nuclei that divides by mitosis. All organisms are composed of eukaryotic cells, except for bacteria. Eukaryotic microorganisms include fungi, algae and protozoa, and prokaryotic microorganisms include the bacteria and blue-green bacteria.

Viruses differ from the rest so significantly that we will omit them from this discussion. Protozoa are single-cell organisms that often have complex and highly ordered structures and are sometimes classified with algae.

Algae are a group of aquatic organisms that contain chlorophyll. They are usually able to survive periods of dryness and are often found in soils, in tree barks and on rocks in coastal or moist climates. They are generally characterized by their colour: green algae inhabit fresh water, while brown algae are marine organisms and are usually found in surface layers

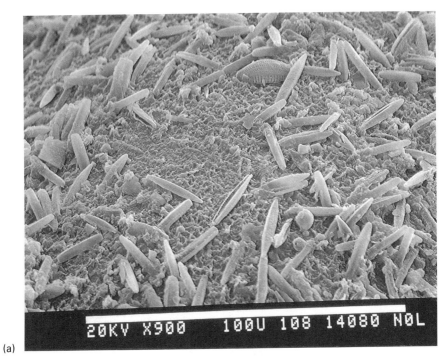

(a)

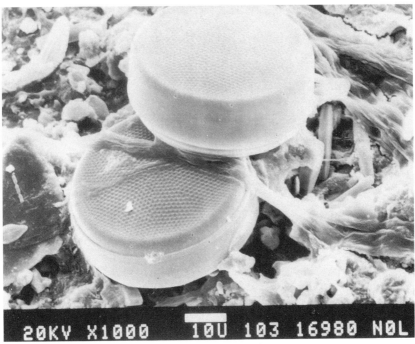

(b)

2.1 SEM micrographs of diatoms from a marine biofilm: (a) pennate (elongate); (b) centric (drum shaped).

of the ocean. Diatoms (Fig. 2.1) are microscopic algae and a source of food for marine life.

Fungi are a group of eukaryotic protists that include yeasts, moulds and mushrooms. They lack chlorophyll and are not photosynthetic. They are often found in soils, and as parasites on living plants and animals.

Bacteria are typically single-celled microorganisms that have no chloro-

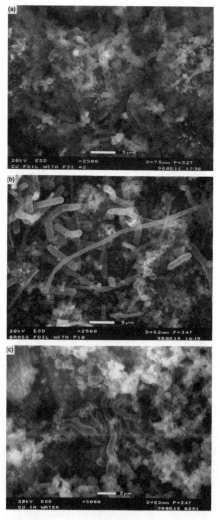

2.2 SEM micrographs of bacterial cells: (a) spherical; (b) rod shaped and filamentous; (c) helical.

phyll, multiply by simple division, and can be seen only with a microscope (Fig. 2.2).

2.2.2   Characteristics of microorganisms

There are primarily three types of characteristics used to classify microorganisms: morphology; physiology; and genetic. While this section will not present an extensive review of microbiology, it will cover some broad classifications, based on microorganisms' metabolic capabilities.[8]

Microorganisms can adhere to almost any surface in contact with natural waters. Organisms reproduce and many produce exopolymers, secretions which form a matrix of fibres.[9, 10] The exopolymers, also called extracellular polymers, influence the surface chemistry on which they are attached. One common example of an exopolymer is snail slime. (See Section 2.4.1 for more information on extracellular polymers.)

Since a consortium of microorganisms is apparently involved in the formation of biofilms, several different types of microbes must, at least temporarily, live together as a small unit.[11, 12]

Oxygen and microbial growth

One common method of grouping bacteria is based on their oxygen requirement. Are they capable of growth in the presence or absence of air or oxygen? Louis Pasteur became aware of the capability of growth relating to oxygen. Before his research into the production of alcohol by yeast, it was thought that life was possible only in the presence of air. We now know otherwise. In fact, the most fundamental classification of microbes is based on their response to air: aerobic or anaerobic organisms. (Aerobic organisms are oxygen-utilizing, while anaerobic organisms tolerate no oxygen.)

These organisms are further classified as strict or obligate aerobes, when they require oxygen, or as strict or obligate anaerobes when they cannot grow in oxygen. Some organisms can grow in either the presence or absence of oxygen, and are known as facultative anaerobes.[13] Microaerophilic organisms, while ill-defined, require oxygen but only in concentrations less (2–10%) than the concentration in air (20%).

Aerobes require oxygen primarily as an electron acceptor for the electron transport system necessary for generating energy. Sometimes an aerobe also requires a small amount of oxygen for selected enzyme reactions. The availability of oxygen in the environment is a central factor in the natural selection of organisms that inhabit an environment. Organisms with different oxygen needs usually grow with each other. Figure 2.3 illustrates this.[14]

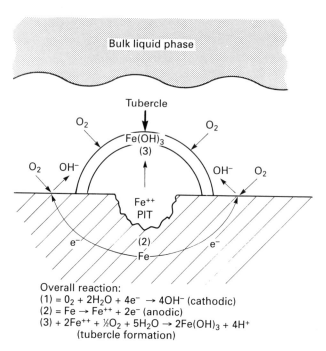

Overall reaction:
(1) = $O_2 + 2H_2O + 4e^- \rightarrow 4OH^-$ (cathodic)
(2) = $Fe \rightarrow Fe^{++} + 2e^-$ (anodic)
(3) + $2Fe^{++} + \frac{1}{2}O_2 + 5H_2O \rightarrow 2Fe(OH)_3 + 4H^+$
      (tubercle formation)

2.3 Differential aeration cell formed by oxygen depletion under a microbial surface film (after Iverson, W.P., 1974, 'Microbial corrosion of iron', in *Microbial Iron Metabolism*, Neilands, J.E., ed., Academic Press, New York, p. 475).

Biochemical oxygen demand

Organic matter oxidizes as it decays. Figure 2.4 is an illustration of the aerobic decay of organic matter.[15] At each step, a part of the process oxidizes to produce CO_2 and H_2O, as well as new organic matter.

The biochemical oxygen demand (BOD) test measures the amount of O_2 used in the aerobic biological decay of organic matter. The BOD test is useful, not to assess the strength of wastewater, but to understand how much oxygen would be used if wastewater were diluted into fresh water.[16] Organisms in the water use the organic matter as a food source. Since the BOD test measures the organic compounds, monitoring the BOD provides a tool for wastewater treatment and control of organic compounds that act as pollutants.

Nutrition and microbial growth

Water makes up most of the weight of a microorganism. The composition shown for the bacterium *Escherichia coli* in Table 2.1 is representative of

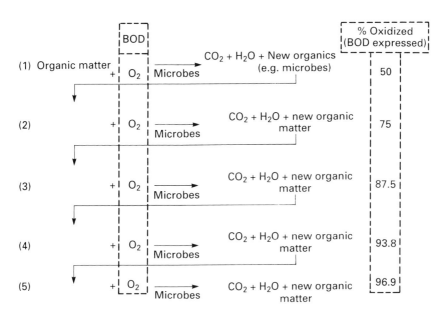

2.4 Aerobic decay: repetitive biological processing of organic matter. The oxygen utilized in this process up to any step is a measure of the biochemical oxygen demand of the original sample of organic matter. If the biological cell yield is 50%, then more than 95% of the original organic matter is totally oxidized by the end of the fifth cycle (figure is from *Elements of Bioenvironmental Engineering* by Gaudy and Gaudy, and is used by permission of Engineering Press, Inc. the copyright owner).

most species of microorganisms.[17] Four elements, carbon, nitrogen, oxygen and hydrogen, make up 90% of the dry weight of the cell.[18] The balance is made up of trace elements. Of these four elements, hydrogen and oxygen are derived from the water used by the cell, while carbon, oxygen and nitrogen are the limiting specific nutritional requirements for the cell.[19]

The three most important nutritional requirements are a carbon source, an energy source and an electron donor.[20] These sources are determined by the enzymic make-up of the cell. Gaudy describes these broad metabolic groups, as shown in Table 2.2.[21] The wide variety of microbial metabolisms makes it difficult to assign microbial species to a single metabolic category.[21] Such assigning is usually done by discussing the microorganisms' growth habits, referring to the metabolic groups rather than to species.

Table 2.1. Elemental cell composition[17]

Element	Dry weight (%)
Carbon	50
Oxygen	20
Nitrogen	14
Hydrogen	8
Phosphorus	3
Sulphur	1
Potassium	1
Sodium	1
Calcium	0.5
Magnesium	0.5
Chlorine	0.5
Iron	0.2
All others	0.3

2.2.3   Habitats of microorganisms

Because of their infinitesimal size, microorganisms are easily carried by air, water and objects. Different types of microorganisms are continuously mixed in the environment, where the right conditions allow them to grow and propagate.[22] Organisms survive because they can cope with different physical and nutritional conditions. As Gaudy discusses, even a small difference, conferring only a slight advantage, can lead to relatively large variations among different species. Because of the extremely rapid growth rates of which microorganisms are capable, such variations can occur with lightning speed.

Temperature and microbial growth

Three temperature ranges – minimum, maximum and optimum – allow the growth of microorganisms although the minimum and maximum temperatures given for a specific organism may be misleading. For example, as shown in Fig. 2.5, at the minimum temperature the growth rate is slow. The rate may increase with increasing temperature, and fall abruptly to zero a few degrees above the optimum.[23]

If the rate of temperature change leading to these temperatures is sufficiently slow, the organism may adapt, and appear outside its range. Such adaptability is significant for industrial situations. According to Gaudy and Gaudy, an organism's temperature values may be altered by other environmental factors, such as pH, concentration of salts or the availability of nutrients.[23]

Microorganisms are characterized by their attraction to temperature for their growth. 'Thermophilic' describes an organism that is attracted to

Table 2.2. Broad metabolic groups[21]

Carbon source	
Inorganic	Autotroph
Organic	Heterotroph
Energy source	
Chemical	Chemotroph
Light	Phototroph
Electron donor	
Inorganic	Lithotroph
Organic	Organotroph

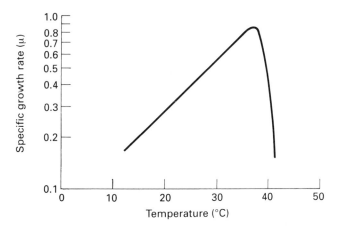

2.5   Hypothetical plot of the relationship between growth rate μ and temperature for a mesophilic bacterium with an optimum temperature for growth of 37 °C (99 °F) and a Q_{10} value of 2. Q_{10} values may not be constant over the entire temperature range (figure is from *Elements of Bioenvironmental Engineering* by Gaudy and Gaudy, and is used by permission of Engineering Press, Inc. the copyright owner).

heat, above 45 °C (110 °F), while 'psychrophilic' describes an organism that is attracted to cold, below 20 °C (70 °F). 'Mesophilic' describes an organism intermediately attracted to heat, 20–45 °C (70–75 °F).

   Growth and survival are not identical, nor are they affected identically by temperature. According to Gaudy and Gaudy, temperatures above the maximum are often lethal and affect both growth and the viability of the organism. Temperatures below the minimum, at which growth is possible, are not usually lethal, however, and so affect growth but not viability.

Microorganisms can stay dormant for very long periods of time with no ill effects.

Dormancy is also relevant to industrial situations. Consider the example of cold groundwater. This water may be used successfully if piped directly into a plant and used quickly. If used in a piping system that runs outside, in which the water is exposed to heat and stays stagnant, the water may develop a clogging problem, or corrode internal pipe surfaces.

Growth and pH

The pH is another environmental factor that influences, and can limit, the microorganism's growth rate. The pH is an indicator of the hydrogen ion concentration (see Section 4.3 for more on pH), which reflects the acidity or alkalinity of an environment. The pH is defined as the negative logarithm of the molar concentration of the hydrogen ion, H^+.[24]

Just as with temperature, microbial species are grouped by growth characteristics of their minimum and maximum pH values. The optimum pH value is that which allows for the optimum growth rate. Gaudy and Gaudy provide several pH preferences of microorganisms:[25]

- Most bacteria have pH optima near neutrality and minimum and maximum pH values for growth near 5 and 9, respectively.
- Most fungi prefer an acid environment and have minimum pH values between 1 and 3 with an optimum pH near 5.
- Most blue-green bacteria have pH optima higher than 7.
- Most protozoa are able to grow in the pH range of 5–8, with an optimum pH near 7.

There are exceptions. The sulphur-oxidizing bacterium *Thiobacillus thiooxidans* grows rapidly in the range 2.0–3.5, and it has been reportedly grown in almost the entire range of less than 0.5 to greater than 9.5, although the practical maximum pH for this organism is around 3.5–4.0.[25]

Changes in pH are often the result of the microorganisms themselves. As Gaudy and Gaudy point out, the internal pH of the cell, unlike the temperature, is not completely determined by the environment.[25] The microorganism can control the passage of ions into and out of the cell. When this ion is hydrogen, the effect of pH on the cell is indirect.

Nutritional requirements

A medium is the specific nutrient solution designed for optimum growth of an organism. Knowing the exact chemical composition of a specific microorganism would be helpful for designing an ideal media for growth. This information is usually not known for the industrial microorganisms, however, although media for medical microbiology are fairly well defined.

Microorganisms cannot reproduce if essential elements are missing in the growth media. A microbiologist has described it as 'trying to get exotic tropical fish to grow in chicken soup'.

Surface environment and microbial growth

Environmental conditions affect the attachment of organisms. These conditions include the water chemistry, temperature, substratum (the base on which an organism lives) and surface roughness of the substrate.

The property of attachment of organisms to a surface is known as adhesion. To separate the factors that influence corrosion, microbiologists measure the chemical characteristics of the biomass, as well as those of the exopolymers.[26]

Eh and electron transport

Eh, the oxidation and reduction (redox) potential, is the measure of the tendency of a solution to give up or accept electrons (or hydrogen or oxygen). *Eh* is the measure of the ability to be oxidized or reduced.[27]

The measurements of *Eh* are expressed as units of potential difference in volts: the more positive the number, the higher the concentration of oxidant to reactant in solution.

Anaerobic microorganisms often grow better if chemicals such as sodium thioglycolate or ascorbic acid are added to keep the medium reduced. Redox indicators such as resusurin may also be added to verify reducing conditions.

Biological oxidation

Biological oxidation is the basis of wastewater treatment systems and operations since it is an easy way to convert soluble organic matter to insoluble matter. Soluble material is metabolized by microorganisms and converted, usually to carbon dioxide and bacterial floc.[28]

An overview of the factors involved in biological oxidation is given in Table 2.3.

2.2.4   General characteristics

While generalizations are helpful in understanding the typical characteristics of microorganisms, in specific situations these generalizations can be misleading. Typical characteristics of MIC-related problem areas include colour, the shape of deposits or penetrations, favourable locations, and the smell and feel to the touch (such as slimy).

Deposits often have a distinct odour and colour. Deposits that are wet

Table 2.3. Factors involved in biological oxidation[28]

Food, BOD	To maintain control with efficient BOD removal
Dissolved oxygen	Insufficient oxygen levels inhibit BOD
pH	With time, bacteria adapt to changes in conditions. Rapid pH changes inhibit the process
Time	The degree of degradation varies with time
Temperature	Low temperatures slow reaction rates; higher temperatures kill some bacteria
Nutrients	Bacteria require nitrogen and phosphorus for cell maintenance

and fresh frequently feel slimy. If iron sulphide (FeS) is present, the sulphur smell may be distinguished from that of hydrogen sulphide.

The bacteria mounds of iron bacteria often feel very fragile, and dissimilar to oxides and mineral deposits. McNeil *et al.* were able to determine certain mineral compositions on copper alloys in conditions that could be formed only by microorganisms.[29] They showed differences in the sulphide films.

Pope provides a list of general characteristics of microorganisms as they relate to MIC:[26]

- They are small, often less than a micrometer (μm).
- They are ubiquitous.
- They are motile as well as sessile.
- They can attach to metal surfaces as needed. Nutrients in water are scarce. Metal surfaces adsorb chemicals and the organisms are attracted to the nutrients they need in the chemicals.
- They are resistant to or tolerate a wide range of temperatures, pressures, pHs and oxygen concentrations.
- They grow in mixed colonies. They can form consortia creating an environment for survival of many organisms that could not survive alone.
- They reproduce rapidly under certain favourable conditions.
- They are resistant to many chemicals.
- They may produce a wide variety of acids, such as acetic, sulphuric and formic.
- They may produce films, known as extracellular polymers. These films, similar to the slime formed by snails, trap nutrients and often act as crevices, or create oxygen concentration cells.
- They may oxidize or reduce metals or metallic ions.

2.3   Microorganisms associated with MIC

As noted above, microorganisms reproduce rapidly and tolerate a vast range of environments. Microorganisms are thought to be primarily responsible for many metal corrosion failures, and are classified by metabolic group, as shown in Table 2.2. These microorganisms are called chemotrophs (divided by energy source) which means they get their energy from a chemical source rather than light (phototroph). They are:

- Sulphate-reducing bacteria.
- Iron and manganese bacteria.
- Sulphur-oxidizing bacteria.

Another method of classification is by oxygen use, such as aerobic or anaerobic, with overlap between the two classifications. In addition, there are the commonly discussed classifications: slime-formers and acid producers.

2.3.1 Sulphate-reducing bacteria

Sulphate-reducing bacteria (SRBs) reduce sulphate to sulphide. Often corrosion to iron and steel occurs under anaerobic conditions in the presence of SRBs. The SRBs use the sulphate ion to produce hydrogen sulphide. Many investigations involving MIC focus on SRBs.[11, 12, 26, 30–38]

Common SRBs include *Desulfovibrio*, *Desulfobacter* and *Desulformaculum*. They are thought to be the dominant and most important bacteria associated with corrosion, and are anaerobic. Until recently, because they were thought to be the primary cause of MIC problems, most research was focused on SRBs. Now we realize that a consortium of microorganisms influences corrosion, and is the most damaging.

In 1934, von Wolzogen Kuhr and Van der Vlugt published a paper about microorganisms in the soil producing corrosion to buried steel pipe.[39] The reactions proposed were:[38–41]

$$4Fe \rightarrow 4Fe^{+2} + 8e^- \qquad [2.1]$$
(anodic reaction)

$$8H_2O \rightarrow 8H^+ + OH^- \qquad [2.2]$$
(dissociation of water)

$$8H^+ + 8e^- \rightarrow 8H \text{ (ads)} \qquad [2.3]$$
(cathodic reaction)

$$SO_4^{-2} + 8H \rightarrow S^{-2} + 4H_2O \qquad [2.4]$$
(depolarization by SRBs)

$$Fe^{2+} + S^{2-} \rightarrow FeS \qquad [2.5]$$
(corrosion product)

$$3Fe^{2+} + 6OH^- \rightarrow 3Fe(OH)_2 \qquad [2.6]$$
(corrosion product)

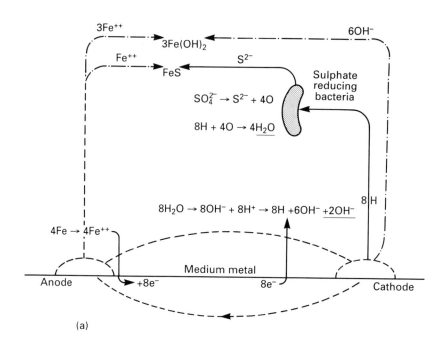

(a)

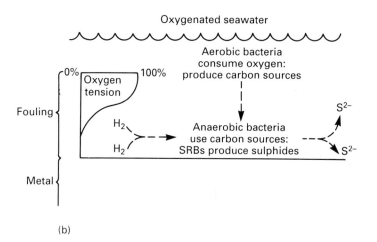

(b)

2.6 (a) Influence of sulphate reducing bacteria on corrosion of iron (source: Sharpley, J.M., 1961, 'Microbiological corrosion in waterfloods' *Corrosion* **17**(8) 386, reproduced with permission); (b) variations through the thickness of a bacterial film. Aerobic organisms near the outer surface of the film consume oxygen and create a suitable habitat for the sulphate reducing bacteria at the metal surface (source: The Metals Society, now The Institute of Materials. National Physical Laboratory, 1983, *Microbial Corrosion*, Proceedings of the Conference).

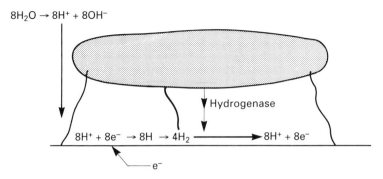

$8H_2O \rightarrow 8H^+ + 8OH^-$

Hydrogenase

$8H^+ + 8e^- \rightarrow 8H \rightarrow 4H_2 \longrightarrow 8H^+ + 8e^-$

e^-

2.7 Cathodic depolarization of surface due to utilization of hydrogen by hydrogenase producing microorganisms (after Fleming, H.C., and Geesey, G.G., ed., 1991, *Biofouling and Biocorrosion in Industrial Water Systems*, Springer-Verlag, Heidelberg, reproduced with permission).

$$4Fe + SO_4^{2-} + 4H_2O \rightarrow 3Fe(OH)_2 + FeS + 2OH^- \qquad [2.7]$$
(overall reaction)

The process is shown graphically in Fig. 2.6.[38, 42] This classic example, originally proposed for a corrosion failure, involved the removal of hydrogen by the bacteria's hydrogenase enzyme. The removal of hydrogen was used by the bacteria to reduce sulphate to sulphide. Without the bacteria, the process stops at equation 2.4, because the surface would be covered by a layer of hydrogen. The bacteria are thought to strip away the hydrogen, a process called cathodic depolarization (Fig. 2.7).[10] (Cathodic depolarization is discussed more fully in Section 2.4.3.)

As the example above indicates, SRBs are associated with the process whereby the sulphate is reduced to sulphur, and the sulphur then reacts with available hydrogen and iron to form hydrogen sulphide and iron sulphide. The result is an alkaline environment. Any metal surface that cannot take this environment will corrode.

Equation 2.7 attempts to describe the role of bacteria in the overall corrosion reaction. Various theories include:

- The importance or necessity of the enzyme hydrogenase to the process.
- The depolarizing influence of iron sulphide.
- Interaction between iron sulphide and hydrogenase.
- The effect of hydrogen sulphide.
- A corrosive metabolite.

SRBs are anaerobic, capable of growing in soil, fresh water or salt water.[43] They can grow in water trapped in stagnant areas, such as dead legs of piping. Because, as noted earlier, some microorganisms can be isolated from their natural environment and can adapt to a new environment, some freshwater organisms may grow in salt water and vice versa.[43]

Many species of SRBs are recognized. They may differ in their microscopic appearance (morphology) or in the substances they metabolize. In general, they all oxidize organic substances to organic acids or CO_2, by the reduction of sulphate to sulphide, by anaerobic respiration.

SRBs tolerate a wide range of pH values, for example, pH of 5–9.5. Since the pH is measured on a log scale, moving from 5 to 6 is a factor of 10; to 7 is 100, to 8 is 1000, and so on.

Black deposits are typical characteristics of SRBs on stainless steel. This black deposit is usually predominantly iron sulphide, and is sometimes a black deposit at the interface between the mound and the metal, with a lighter brown deposit on top. There may be a pit with black FeS at the centre and rings, sometimes bluish, around them. Austenitic stainless steels rely on a passive film for protection against corrosion. This film is primarily Cr_2O_3.

Sanders and Hamilton confirmed microbial corrosion due to SRBs in North Sea oil production systems.[32] They identified SRBs, sulphur-oxidizing bacteria, hydrocarbon-oxidizing bacteria, iron-oxidizing bacteria, slime-forming bacteria and fungi.

In summary, the mechanisms by which sulphate reducers corrode metal is not agreed upon, but there are several possibilities.[44] Primarily, SRBs accelerate corrosion by:[45]

- Generating H_2S.
- Creating oxygen concentration cells.
- Forming insoluble sulphides when metal ions combine with sulphur.
- Cathodically depolarizing.

Bibb and Hartman and Pope *et al.* reviewed various theories to describe the role of SRBs in the corrosion process.[33, 46] Licina summarizes SRB mechanisms as:[44]

- The evidence for cathodic depolarization (whatever the exact mechanism) is quite convincing.
- The precise roles of hydrogenase, ferrous ion, iron sulphide and hydrogen sulphide have not been positively determined, but are probably interdependent upon the local environmental and ecological conditions.
- The corrosive metabolite theory is interesting but not well established. A correlation between MIC and the increased pollution of surface waters with phosphate, nitrate and sulphate over the last 15 years is circumstantial evidence for such a theory.

2.3.2   Iron or manganese bacteria

Iron-oxidizing bacteria, such as *Gallionella*, *Sphaerotilus*, *Leptothrix* and *Crenothrix*, oxidize iron from ferrous (iron II) iron, a soluble form of iron, to ferric (iron III) iron, an insoluble form of iron (Fig. 2.8).[10]

Gallionella are stalk-forming bacteria, are obligately aerobic, and exude metal oxide. They produce insoluble waste high in iron and manganese. They seem to concentrate chloride ions, which aggravates corrosion, particularly in crevice conditions and in austenitic stainless steels. Pits in stainless steel are generally very small at the surface, with large cavities subsurface.[11, 26, 31, 47, 48]

The iron and manganese bacteria may be aerobic and can oxidize ferrous iron to ferric iron, which can attract the chloride ion and produce ferric chloride (Fig. 2.9).[10] Ferric chloride is extremely aggressive and pits stainless steel. The deposits from the bacteria are rich in ferric chloride, and deposits aggressively attack austenitic stainless steel. Ferric chloride produces a very low pH (about 1).[49] While ferric chloride is very aggressive on stainless steel, it can also corrode carbon steel, albeit usually at a slower rate.

Sphaerotilus, *Leptothrix*, *Clonothrix* and *Crenothrix* are aerobic, fila-

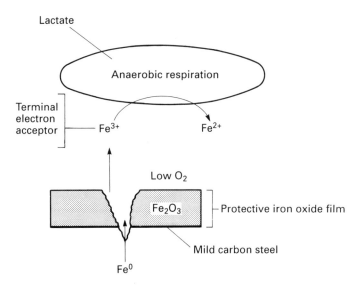

2.8   Iron production by anaerobic respiration of ferric iron associated with protective gamma ferric oxide film on mild carbon steel (after Fleming, H.C., and Geesey, G.G., ed., 1981, *Biofouling and Biocorrosion in Industrial Water Systems*, Springer-Verlag, Heidelberg, reproduced with permission).

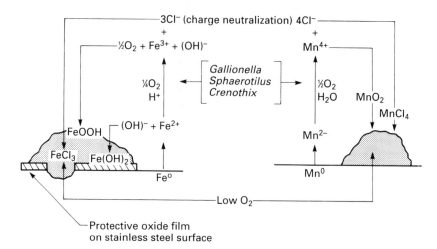

2.9 Iron and manganese oxidation and precipitation in presence of filamentous bacteria. Stainless steel pitting in presence of chloride ions concentrated at surface in response to charge neutralization of ferric and manganic cations (after Fleming, H.C., and Geesey, G.G., ed., 1991, *Biofouling and Biocorrosion in Industrial Water Systems*, Springer-Verlag, Heidelberg, reproduced with permission).

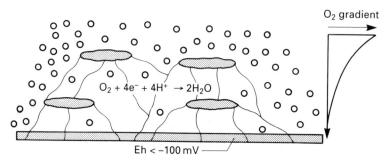

2.10 Oxygen concentration gradient in biofilm caused by respiratory activity of microorganisms (after Fleming, H.C., and Geesey, G.G., ed., 1991, *Biofouling and Biocorrosion in Industrial Water Systems*, Springer-Verlag, Heidelberg, reproduced with permission).

mentous bacteria that can affect both stainless steel and carbon steel.[47] They oxidize soluble Fe to insoluble $Fe(OH)_3$. They may also concentrate and oxidize manganese.[31, 50]

Pseudomonas is an aerobic slime-former and often forms thin films

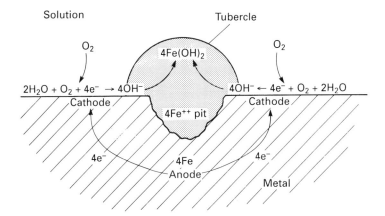

2.11 Oxygen concentration under tubercle (after Tiller, A.K., 1986, 'A review of the European research effort on microbial corrosion between 1950 and 1984', in *Biologically Induced Corrosion*, Dexter, S.C., ed., NACE, Houston, TX, reproduced with permission).

combined with corrosion deposits on metal surfaces.[51] It scavenges oxygen and harbours other species.[31] It reduces iron; this reaction of Fe^{3+} to Fe^{2+} seems to expose active metal surfaces and corrode metal.

Oxygen concentration cells will be discussed in Chapter 4, but a simplified illustration in Fig. 2.10 shows how tubercles can form and Fig. 2.11 shows how pits can grow.[10,35] Pits grow as ferrous ions collect at the base of the pit and chloride ions concentrate (to stay electrically neutral). Pit growth is a two-step process: initiation and propagation. The microbes form a film and an oxygen concentration cell initiates a pit. Localized corrosion drives the pit to propagate deeper. The pit growth is unaffected by the presence of microbes.

For iron bacteria on austenitic stainless steel, the deposits are typically brown or red-brown mounds. There may be rust-coloured streaks on the surface, if the surface is vertical, which may run both up and down from the affected region marked by pits.[11] Low, cone-shaped mounds sometimes appear on top of the pits: the iron-oxidizing bacteria oxidize the ferrous ion to the insoluble ferric state; the resulting deposits create a mound.

Often the pits and mounds are at low points of tanks or pipes, probably because of remaining stagnant water – an environment suitable for growth of the microorganisms. Mounds and pitting may appear at areas where condensation collects, in interfaces, oil–water lines, air–water splash zones, or in fuel oil, areas of gasoline and fuel.[52,53]

2.3.3   Sulphur-oxidizing bacteria

Sulphur-oxidizing bacteria are aerobic bacteria. They oxidize elemental sulphur or sulphur-bearing compounds, producing sulphuric acid (Fig. 2.12).[54,55] Sulphuric acid is corrosive to many metals.

Gaudy and Gaudy discuss the aerobic bacteria that oxidize sulphur, called colourless sulphur bacteria, which distinguishes them from the oxidizing photosynthetic bacteria.[56] The colourless sulphur bacteria include four genera, *Thiobacillus*, *Thiodendron*, *Beggiatoa* and *Sulfolobus*, probably the most studied of the colourless sulphur bacteria. Concrete installations and pipe are often rapidly destroyed by the growth of *Thiobacillus*.[57]

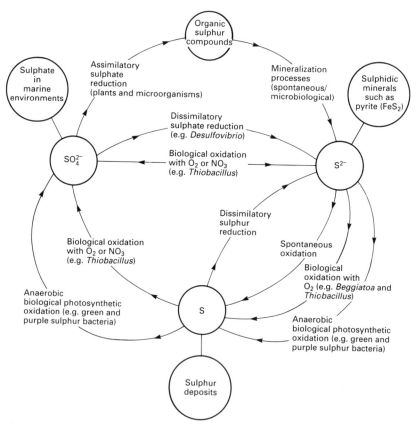

2.12   The sulphur cycle showing the role of bacteria in oxidizing elemental sulphur to sulphate (SO_4^{2-}) and in reducing sulphate to sulphide (S^{2-}) (source: National Physical Laboratory, 1983, *Microbial Corrosion, Proceedings of the Conference*, The Metals Society, now The Institute of Materials, UK, 1983).

One species, *T. ferroxidans* is of particular interest, as it is able to oxidize both ferrous iron and sulphur compounds. *T. ferroxidans*, as well as *T. thiooxidans*, is capable of growth under very acid pH values (1.0 or less).

2.3.4 Other methods of classifying microorganisms associated with MIC

Microorganisms, while commonly classified by oxygen use, can also be classified as slime-formers and acid producers.

Aerobic bacteria

Biofilms, which form under aerated conditions, make a type of corrosion called differential aeration cell or concentration cell (see Section 1.2 for more details). Aerobic corrosion occurs when the oxide film is damaged, or oxygen is kept from the metal surface by the microorganisms.[58] Rapid pitting or corrosion occurs under the biofilm.

An example of aerobic corrosion occurs when an iron-oxidizing bacteria (an aerobic bacteria such as *Gallionella*) accumulates on a steel surface. A tubercle forms and corrosion occurs underneath, which in turn encourages further corrosion. Aggressive anions such as chlorides accumulate in the pit. These anions drive the pit deeper and reduce the pH. The iron-oxidizing bacteria oxidize soluble iron and manganese and deposit ferric and manganic chlorides.[31, 58]

This form of corrosion is common in steels as well as stainless steels. Stainless steels rely on a stable oxide film to provide corrosion resistance (see Section 4.5).

Anaerobic bacteria

Anaerobic corrosion is associated with bacteria that grow under anaerobic conditions.[58] Inner layers of microorganisms in the biofilm are usually anaerobic.

SRBs are obligate anaerobes that can function only in the absence of oxygen. Their make-up means they must use hydrogen for growth and sulphate or other reduced-sulphur compounds as the terminal electron acceptor.[59] Sulphide is their metabolic product (see Section 2.3.1). See Fig. 2.13 for a schematic representation of the two mechanisms.[59–61]

Acid-producing bacteria

Bacteria can release aggressive metabolites, such as organic (acetic, succinic, icobuteric, etc.) or inorganic acids (sulphuric).[59] Corrosion

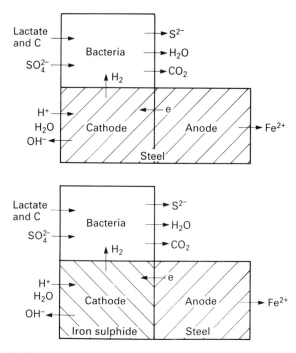

2.13   Schematic representation of the classical mechanism (top) and alternative mechanism (bottom) of microbial corrosion under anaerobic conditions (reproduced with permission from *Nature*, 1971, King, R.A., and Miller, J.D.A., 'Corrosion by the sulfate reducing bacteria', **233** 491. Copyright from Macmillan Magazines Limited).

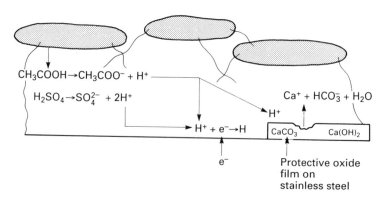

2.14   Acid production (organic and inorganic) by adherent film-forming bacteria with consequent promotion of electron removal from cathode by hydrogen or dissolution of protective calcareous film on stainless steel surface (after Fleming, H.C., and Geesey, G.G., ed., 1991, *Biofouling and Biocorrosion in Industrial Water Systems*, Springer-Verlag, Heidelberg, reproduced with permission).

occurs underneach these bacteria, with underdeposit corrosion caused by acid-producing bacteria (APB), as shown in Fig. 2.14.[10]

For instance, *Thiobacillus thiooxidans* produce H_2SO_4. The corrosion proceeds, and hydrogen is liberated as the cathodic reaction.[62–64] Tiller gives the typical reaction for this sulphur-oxidizing bacteria (SOB) as:[59]

$$H_2S + 2O_2 \rightarrow H_2SO_4 \qquad [2.8]$$

$$2S + 3O_2 + 2H_2O \rightarrow 2H_2SO_4 \qquad [2.9]$$

$$5Na_2S_2O_3 + 4O_2 + H_2O \rightarrow 5Na_2SO_4 + H_2SO_4 + 4S \qquad [2.10]$$

Little *et al.* showed that under laboratory conditions an aerobic acetic APB can accelerate the corrosion of cathodically protected stainless steel in synthetic salt solution.[65] The acetic acid destabilizes or dissolves the calcareous film that formed during cathodic polarization.

Slime-formers

Slime-formers, as their name so subtly implies, form slime. Iron bacteria commonly plug water systems and produce slimy precipitates of hydrated ferric oxide.[42, 59, 66, 67]

Other bacteria are grouped together and are known collectively as slime-forming bacteria. These include *Pseudomonas*, *Escherichia*, *Flavobacterium*, *Aerobacter* and *Bacillus*.[68, 69] They grow in a patchy distribution over the metal surface and occlude oxygen via respiration; the slime impedes oxygen diffusion, creating an oxygen concentration cell.[70] The slime plays a role in the aggregation of bacteria in flocs and in films on surfaces. Slime layers influence corrosion as discussed further in Section 2.4.1.

2.3.5 Other characteristics and features

Researchers also classify microorganisms by the influence of differential aeration cells, metabolic processes of the microorganisms, and by how the microorganisms influence anodic and cathodic reactions.

2.4 Metabolic processes

Metabolic reactions are chemical reactions in living organisms.[1, 35] In many microbes, these reactions involve oxidation–reduction processes accompanied by the release of energy.[1] Since microbes may act as electron donors or electron acceptors, microorganisms attached to metal surfaces are capable of producing an electrochemical reaction.

There are many proposed mechanisms (automatic and consistent

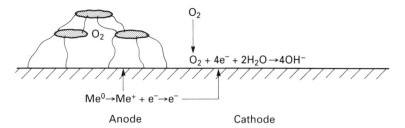

2.15   Differential aeration cell resulting from heterogeneous distribution of biofilm-forming microorganisms over the surface (after Fleming, H.C., and Geesey, G.G., ed., 1991, *Biofouling and Biocorrosion in Industrial Water Systems*, Springer-Verlag, Heidelberg, reproduced with permission).

responses of an organism to various stimuli) for MIC, based on how microorganisms influence corrosion. For example, layers of oxygen-deficient conditions exist within the biofilm, because oxygen is taken up by respiring (aerobic) organisms near the surface of the film as quickly as the oxygen can diffuse into the film. This situation creates potential differences (the voltage difference between two points) between the metal surface under the biofilm and the surrounding metal surface.[71–73]

A differential aeration cell (Fig. 2.15) is set up between the biofilm underlying the metal surface and the surrounding metal surface.[10] Corrosion is accelerated under the deposit because of the depleted oxygen, as shown in Fig. 2.16 (explained in more detail in Section 4.5.2). This mechanism (oxygen utilization) is based on an electron charge transfer in which oxygen is consumed as an electron charge acceptor (where it is available on the metal surface away from the biofilm).

A second, similar mechanism results with anaerobic bacteria. The differential aeration cells accelerate the bacterial growth. Instead of oxygen, the bacteria will use some other mechanism (such as sulphates or nitrates) as the electron acceptors in the zone under the biofilm (although the corrosion is again accelerated because of depleted oxygen). As noted in Section 2.2.2, bacteria can change. They grow in the presence or absence of an environmental factor such as oxygen; such adaptability is termed a facultative response. Thus, an aerobic bacteria that depletes the supply of oxygen and then begins to use another electron acceptor is defined as a facultative aerobe.[74]

2.4.1   Mechanical influences

In addition to microbiological influences, mechanical influences also affect corrosion. In many cases, these influences are associated with colonies of

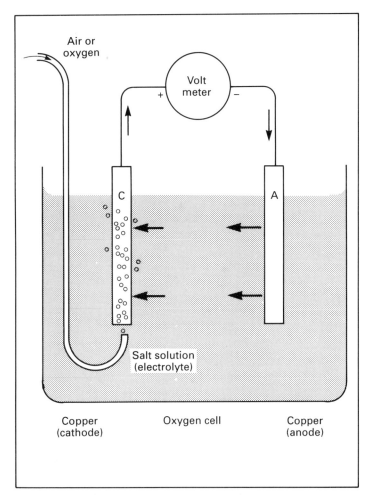

2.16 Laboratory-type corrosion cell resulting from differential oxygen concentration (source: NACE, 1984, *Corrosion Basics, An Introduction*, NACE, Houston, TX, reproduced with permission).

organisms, scale, debris and exopolymers forming local deposits on a metal surface. The deposits often initiate pits underneath the surface of the metal. The pits grow and propagate, leading to corrosion of the metal.

Slime layers produce crevices

Slime layers can influence corrosion. Microscopic and macroscopic organisms are thought to influence corrosion in two ways:[75]

- By creating mats or obstructions on the surface which produce differential aeration cells, which in turn form crevices, under which corrosion is accelerated.
- By absorbing hydrogen from the surface of the steel, which removes hydrogen and accelerates the corrosion process.

Exopolymers

Exopolymers, or extracellular polymers, are part of the biofilm and are thought to be key to the corrosive nature of the corrosion cell. They position bacteria and their enzymes at solid surfaces.[76]

Surfaces in natural waters without toxins or biocides rapidly become covered by biofilms.[66, 77] As discussed in Section 1.2, biofilm develops in several stages:[66]

- Dissolved organic compounds are adsorbed onto the surface to begin the film development. This action may change the surface charge, the contact angle and other properties on the surface.[78, 79] When more fouling occurs, organic compounds may or may not be attracted to a modified surface.
- Bacteria arrive, adhere to the conditioned surface and reproduce. A typical developing biofilm is shown in Fig. 2.17.
- Interactions occur between the bacteria that settle, and the mucus-like biofilm develops. Extracellular polymers or exopolymers have been described as a polymeric matrix. The bacteria may be only about 10% of the total weight, with the rest of the biofilm being exopolymers.[80] When bacteria are in a film, they are very resistant to biocides. In fact, they often produce more exopolymers after biocide treatment to protect themselves, thus aggravating the corrosion problems.
- Organisms other than bacteria arrive and develop, depending on the environment, flow rate, nutrients, light and so forth. Examples are protozoa and algae, which may or may not have preferences for adhering to a biofilm.
- Organisms and the community grow to the limit of the environment. As the nutrients are exhausted, some of the film dies and sloughs off.

2.4.2   *Chemical influences*

In many cases, chemical influences are associated with acid production or hydrogen sulphide production. The chemicals may deposit on a metal surface. The deposits then often initiate pits underneath the surface of the metal. The pits grow and propagate, leading to corrosion of the metal.

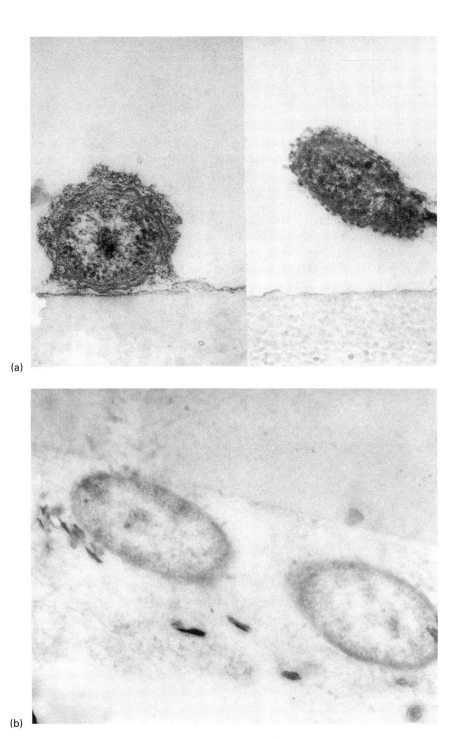

(a)

(b)

2.17   (a) Bacteria adhere to surface; (b) bacteria reproduce.

Acid production

Acid, such as sulphuric acid, is produced as a reaction product in some microbiological processes. Consider the example of slime on underwater pipe. Assume that SRBs are present and reduce the sulphate to sulphides. The sulphides are released into the bulk and are oxidized, assuming the presence of turbulent flow or other process variables, through which sufficient oxygen would be available. If there is not enough oxygen, the sulphide may be combined as hydrogen sulphide gas. This now meets a moist layer on the metal surface, such as with the presence of an SOB such as *Thiobacillus*. This situation, especially common in sewer systems, produces sulphuric acid (H_2SO_4) and aggressively corrodes pipe, as well as ceramics and concrete. Sulphuric acid is a mineral acid with a very low pH. Other microorganisms produce organic acids that are similarly corrosive. Figure 2.6 shows a diagram of this process.

Acids also assist in the breakdown of nonmetallic coatings. The coatings (discussed in Section 7.1) act as a barrier between a corrosive environment and the metal (or material) surface. When breakdown occurs, in the case of MIC to a barrier coating on steel, for example, the corrosion is very localized and aggressive.

H_2S production

The corrosion of iron-based alloys in the presence of hydrogen sulphide and water is dependent on what happens to the hydrogen sulphide. H_2S is produced as a reaction product.[81] Consider the following:

Anodic reaction:

$$Fe \rightarrow Fe^{+2} + 2e^-$$

[2.11]

Cathodic reactions:

$$H_2S + H_2O \rightarrow H^+ + HS^- + H_2O$$

[2.12]

$$H_2O + HS^- \rightarrow H^+ + S^{-2} + H_2O$$

[2.13]

producing the reaction:

$$2e^- + 2H^+ + Fe^{+2} + S^{-2} \rightarrow 2H + FeS$$

[2.14]

for a net reaction of:

$$Fe + H_2S \rightarrow FeS + 2H$$

[2.15]

Another example based on SRBs is discussed in Section 2.3.1.[82] This type of corrosion, in which H_2S is produced, is often associated with failures due to sulphate-reducing bacteria of carbon steel, stainless steel, copper and aluminium alloys.

2.4.3 Electrochemical influences

Electrochemical influences include oxygen, cathodic depolarization and metal oxidation states. Corrosion is usually an electrochemical reaction; not only do chemical reactions occur, but so does a transfer of electrons.

Oxygen

Oxygen is a common cathodic depolarizer, as can be demonstrated by placing iron filings into two beakers of water.[83] One beaker is filled with tap water that has been allowed to sit overnight, and one with fresh, commercially available carbonated water. This experiment can also be demonstrated by bubbling oxygen through one beaker and nitrogen through the other. This saturated solution with nitrogen eliminates dissolved oxygen. The iron filings in the oxygen-rich environment immediately begin to corrode, while the iron filings in the oxygen-poor environment remain bright for quite a while.

An oxygen concentration cell, also referred to as a differential aeration cell, is often deemed a galvanic cell, caused by the difference in oxygen concentration at two points on the surface of a metal. Oxygen from the bulk environment is unable to replenish the areas under the biofilm. First, the film acts as a barrier for the diffusion process of oxygen migrating through the film. The oxygen that does get through is then immediately used by the microorganisms in their metabolic process.[84]

Cathodic depolarization

Corrosion slows in response to the anodic and cathodic processes.[85] The cathodic reaction slows down if the hydrogen product of the cathode is not removed, which can be by evolution of a gas, or some reaction involving oxygen. Cathodic polarization is the polarization of the cathode. The initial potential goes down because of current flow effect at or near the cathode. The potential becomes more active (negative) because of cathodic polarization. Polarization at the anode and cathode determine the corrosion rate generated by most electrochemical cells.[85]

Cathodic depolarization is the removal of the cathodic hydrogen, which depolarizes the cell, greatly increasing the corrosion rate. Cathodic depolarization is frequently discussed as a mechanism for MIC (Fig. 2.7 and 2.18).[86, 87] Microorganisms are able to use the hydrogen produced at the cathode, as well to produce depolarizing compounds.

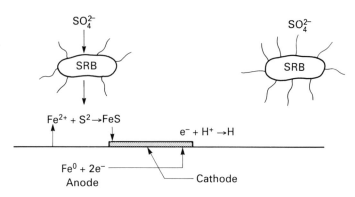

2.18 Cathodic depolarization of surface by iron sulphide as a result of respiration of sulphate by SRBs (after Fleming, H.C., and Geesey, G.G., ed., 1991, *Biofouling and Biocorrosion in Industrial Water Systems*, Springer-Verlag, Heidelberg, reproduced with permission).

Metal oxidation states

An atom of iron metal, for example, is neutral or balanced and is denoted as Fe. If the charge on the nucleus is unbalanced, the atom displays a positive charge denoted as Fe^{2+}. This charged atom is an ion, and the process is ionization. An iron atom that has been stripped of two electrons is Fe^{2+} and is called a ferrous ion. In a similar manner, an iron ion stripped of three electrons is Fe^{3+} and is called a ferric ion. This process of stripping electrons is termed oxidation.

Oxidation can occur in reverse when extra ions are added to a neutral atom, which gives a net negative charge. The increase in negative charge or the decrease in positive charge is called reduction. These particles, the ions, are responsible for carrying current in an aqueous solution.

For example, the ion Fe^{3+} contributes to corrosion as ferric ion corrosion by the process of:

$$2Fe^{3+} + Fe \rightarrow 3Fe^{2+} \hspace{2cm} [2.16]$$

The metal ion Fe^{+3} is very oxidizing.[82] The ferric and ferrous ions behave differently, because of their different metal oxidation states and the fact that they form different compounds. Microorganisms can reduce ferric ions (Fe^{3+}) to ferrous ions (Fe^{2+}) and increase corrosion.

2.5   Microbiological factors affecting MIC

When microorganisms live in communities, the aerobic and anaerobic organisms form layers, and within those layers they can form their own environments; see Fig. 2.1. If a piping system, for example, is in a highly oxygenated environment, you might think that anaerobic organisms would not be able to survive. You would be wrong. Dr D.C. White, a PhD and MD at the University of Tennessee, and director of the Institute of Applied Microbiology, gives an excellent illustration. The lungs and respiratory system are a highly oxygenated system. Microbes attach to our teeth and form communities that we know as plaque. The microbes that cause 'bad breath' are anaerobic bacteria. Clearly, anaerobic bacteria can thrive in an aerobic environment. The same scenario is true for biofilms on pipe in oxygenated systems.

Admittedly, it is difficult to simulate such scenarios of real-life failures in laboratory experiments. Exactly what conditions need to occur for microorganisms to attach and colonize on a metal surface? And what are the precise mechanisms that enable microorganisms to attach and colonize?

Fortunately, a tremendous amount of research on MIC is taking place, involving such disciplines as microbiology, metallurgy, civil engineering, environmental engineering, microbial ecology, corrosion, electrochemistry, chemistry, electrical engineering, water chemistry and surface chemistry. Understanding corrosion involves understanding the interrelationship of all these disciplines. For corrosion to occur in an aqueous environment, four elements are necessary: an anode, a cathode, an

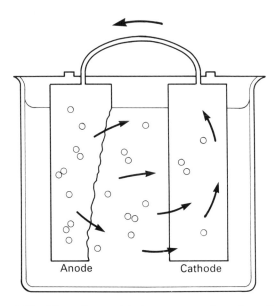

2.19 Flow of current between an anode and a cathode in a corrosion cell (source: NACE, 1984, *Corrosion Basics, An Introduction*, NACE, Houston, TX, reproduced with permission).

electrolyte and a current path (or circuit) as illustrated in Fig. 2.19.[88] The corroding metal (or area) is the anode. The noncorroding metal (or area), discrete from the anode, is the cathode. The electrolyte is the liquid in which the metals are in contact. The electrons (and electricity) flow through the path between the two electrodes, the anode and the cathode. Corrosion will be covered extensively in Chapter 4.

2.5.1   The influence of differential aeration cells

A differential aeration cell consists of an oxygen concentration cell, the result of a potential difference caused by different amounts of oxygen dissolved at two locations.[35, 89] See Fig. 2.20 for a simplified illustration. This is an example of an oxygen concentration cell, or a differential aeration cell that can form under a tubercle. See Sections 2.4.3 and 4.3.2 for more discussion.

Differential aeration cells also provide a condition for sulphate-reducing bacteria, such as *Desulfobacter*, to grow.

Similarly, *Pseudomonas*, *Sphaerotilus* and *Desulfovibrio* have been shown to be associated with the corrosion of stainless steels.[48] Filamentous iron bacteria, such as *Sphaerotilus*, *Crenothrix* and *Leptothrix*, as well as the stalk-forming *Gallionella*, oxidize ferrous (Fe^{2+}) ions to ferric (Fe^{3+}) ions. This oxidation results in thick deposits of ferric hydroxide. A comparable mechanism exists with other bacteria for oxidizing manganous (Mn^{2+}) ions to manganic (Mn^{3+}) ions.

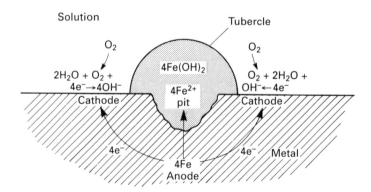

2.20 Schematic of pit initiation and tubercule formation due to an oxygen concentration cell under a biological deposit (source: Tiller, A.K., 1986, 'A review of the European research effort on microbial corrosion between 1950 and 1984', in *Biologically Induced Corrosion*, Dexter, S.C., ed., NACE, Houston, TX, reproduced with permission).

2.5.2   The influence of bacteria's metabolic process

Another corrosion mechanism is based on the by-products of the bacteria's metabolic process (the complex group of physical and chemical processes involved in the maintenance of life). For instance, *Thiobacillus thiooxidans* produces H_2SO_4. Acid metabolites (from bacteria) accelerate corrosion by dissolving oxides from the metal surface and accelerating the cathodic reaction rate.[90–93]

2.5.3   The influence of anodic and cathodic reactions

The bacteria can alter anodic and cathodic reactions depending on the environment and the organisms involved.[94] Equation 2.17 illustrates corrosion as a chemical reaction:

$$Fe + 2HCl \rightarrow FeCl_2 + H_2 \qquad [2.17]$$

This is a metal dissolved by an acid, thereby producing a soluble salt and hydrogen gas.

An electrochemical reaction is a chemical reaction involving the transfer of electrons.[95] This concept of corrosion and electrochemistry is discussed in Chapter 4. It is a chemical reaction which involves oxidation and reduction, the process of removing electrons from atoms being oxidation. For example, Fe is the chemical symbol for a neutral iron atom, and Fe^{++}, or Fe^{2+}, means the iron atom has been stripped of two electrons, and is known as ferrous iron.

The opposite process is termed reduction. In this reaction the atom gains electrons to result in a net negative charge, which means that, in the case of Equation 2.17, the process can be simplified into an oxidation reaction and a reduction reaction.

$$Fe \rightarrow Fe^{++} + 2e^- \qquad [2.18]$$

is an oxidation reaction and occurs at the anode.

$$2H^+ + 2e^- \rightarrow + H_2 \qquad [2.19]$$

is a reduction reaction and occurs at the cathode.

2.5.4   The influence of electron transport in bacteria

Electron transport in bacteria is complex. Some bacteria have the ability to use alternative final electron acceptors in the absence of oxygen, while others, such as the SRBs, can use a variety of electron acceptors at different oxidation levels.[96]

Suggested reading

1 Characklis, W.G., and Marshall, K.C., 1990, *Biofilms*, John Wiley & Sons, New York.
2 Fleming, H.C., and Geesey, G.G., 1991, *Biofouling and Biocorrosion in Industrial Water Systems*, Springer-Verlag, New York.
3 Gaudy, A., and Gaudy, E., 1980, *Microbiology for Environmental Scientists and Engineers*, McGraw-Hill, New York.
4 Pope, D.H., Duquette, D., Wayner, P.C., and Johannes, A.H., 1989, *Microbiologically Influenced Corrosion: A State of the Art Review*, MTI Publication No. 13, National Association of Corrosion Engineers, Houston, TX.
5 Dexter, S.C., ed., 1986, *Biologically Influenced Corrosion*, NACE Reference Book No. 8, National Association of Corrosion Engineers, Houston, TX.
6 Ford, T.E., ed., 1993, *Aquatic Microbiology: An Ecological Approach*, Blackwell, Cambridge, MA.
7 NACE, 1993, *A Practical Manual on Microbiologically Influenced Corrosion*, National Association of Corrosion Engineers, Houston, TX.
8 NACE, 1984, *Corrosion Basics, An Introduction*, National Association of Corrosion Engineers, Houston, TX.
9 Pope, D.H., 1986, *A Study of Microbiologically Influenced Corrosion in Nuclear Power Plants and a Practical Guide for Countermeasures*, EPRI NP-4582, Electric Power Research Institute, Palo Alto, CA.
10 Licina, G.J., 1988, *Sourcebook for Microbiologically Influenced Corrosion in Nuclear Power Plants*, EPRI NP-5580, Electric Power Research Institute, Palo Alto, CA.
11 Licina, G.J., 1988, *Detection and Control of Microbiologically Influenced Corrosion*, EPRI NP-6815-D, Electric Power Research Institute, Palo Alto, CA.
12 Dillon, C.P., 1986, *Corrosion Control in the Chemical Process Industries*, McGraw-Hill, New York.
13 NACE, TPC 3, 1990, *Microbiologically Influenced Corrosion and Biofouling in Oilfield Equipment*, National Association of Corrosion Engineers, Houston, TX.
14 Miller, J.D.A., 1970, *Microbial Aspects of Metallurgy*, American Elsevier, New York.
15 Gaudy, A., and Gaudy, E., 1988, *Elements of Bioenvironmental Engineering*, Engineering Press, San Jose, CA.

References

1 Pope, D.H., Duquette, D., Wayner, P.C., and Johannes, A.H., 1989, *Microbiologically Influenced Corrosion: A State of the Art Review*, MTI Publication No. 13, NACE, Houston, TX.
2 Little, B.J., 1985, 'Succession in Microfouling', *Proceedings of Office of Naval Research on Marine Biodeterioration*.
3 Uhlig, H.H., 1948, *Corrosion Handbook*, John Wiley & Sons, New York, p. 466.

4   Fontana, M., 1986, *Corrosion Engineering*, McGraw-Hill, New York, p. 393.

5   Characklis, W.G., 1984, 'Biofilm development: a process analysis', in *Microbial Adhesion and Aggregation*, Marshall, E.C., ed., Springer-Verlag, New York.

6   Pope, D.H., and Zintel, T.P. 1988, 'Methods for the investigation of underdeposit microbiologically influenced corrosion', in *NACE Corrosion/88*, National Association of Corrosion Engineers, Houston, TX.

7   Pope, D.H., Siebert, O.W., Zintel, T.P., and Kuruvilla, A.K., 1988, 'Organic acid corrosion of carbon steel: A mechanism of microbiologically influenced corrosion', in *NACE Corrosion/88*, National Association of Corrosion Engineers, Houston, TX.

8   Gaudy, A., and Gaudy, E., 1980, *Microbiology for Environmental Scientists and Engineers*, McGraw-Hill, New York, p. 343.

9   Costerton, J.W., and Geesey, G.G., 1986, 'The microbial ecology of surface colonization and of consequent corrosion', in *Biologically Influenced Corrosion*, Dexter, S.C., ed., NACE Reference Book No. 8, National Association of Corrosion Engineers, Houston, TX.

10 Fleming, H.C., and Geesey, G.G., 1991, *Biofouling and Biocorrosion in Industrial Water Systems*, Springer-Verlag, New York.

11 Kobrin, G., 1976, 'Corrosion by microbiological organisms in natural water', *Materials Performance* **15**(17) 38.

12 Hamilton, W.A., and Maxwell, S., 1986, 'Biological and corrosion activities of sulfate-reducing bacteria with natural biofilms', in *Biologically Influenced Corrosion*, Dexter, S.C., ed., NACE Reference Book No. 8, National Association of Corrosion Engineers, Houston, TX.

13 Gaudy, A.F., and Gaudy, E.T., 1980, *Microbiology for Environmental Scientists and Engineers*, McGraw-Hill, New York, p. 188.

14 Iverson, W.P., 1974, 'Microbial corrosion of iron', in *Microbial Iron Metabolism*, Neilands, J.B., ed., Academic Press, New York, p. 475.

15 Gaudy, A.F., and Gaudy, E.T., 1980, *Microbiology for Environmental Scientists and Engineers*, McGraw-Hill, New York, p. 60.

16 Gaudy, A.F., and Gaudy, E.T., 1980, *Microbiology for Environmental Scientists and Engineers*, McGraw-Hill, New York, p. 487.

17 Gaudy, A.F., and Gaudy, E.T., 1980, *Microbiology for Environmental Scientists and Engineers*, McGraw-Hill, New York, p. 59.

18 Gaudy, A.F., and Gaudy, E.T., 1980, *Microbiology for Environmental Scientists and Engineers*, McGraw-Hill, New York, p. 195.

19 Gaudy, A.F., and Gaudy, E.T., 1980, *Microbiology for Environmental Scientists and Engineers*, McGraw-Hill, New York, p. 196.

20 Gaudy, A.F., and Gaudy, E.T., 1980, *Microbiology for Environmental Scientists and Engineers*, McGraw-Hill, New York, p. 344.

21 Gaudy, A.F., and Gaudy, E.T., 1980, *Microbiology for Environmental Scientists and Engineers*, McGraw-Hill, New York, p. 344.

22 Gaudy, A.F., and Gaudy, E.T., 1980, *Microbiology for Environmental Scientists and Engineers*, McGraw-Hill, New York, p. 176.

23 Gaudy, A.F., and Gaudy, E.T., 1980, *Microbiology for Environmental Scientists and Engineers*, McGraw-Hill, New York, p. 178.

24 Gaudy, A.F., and Gaudy, E.T., 1980, *Microbiology for Environmental Scientists and Engineers*, McGraw-Hill, New York, p. 193.

25   Gaudy, A.F., and Gaudy, E.T., 1980, *Microbiology for Environmental Scientists and Engineers*, McGraw-Hill, New York, p. 183.

26   Pope, D.H., 1986, *A Study of Microbiologically Influenced Corrosion in Nuclear Power Plants and a Practical Guide for Countermeasures*, EPRI NP-4582, Electric Power Research Institute, Palo Alto, CA.

27   ASM, 1981, *Manual for Methods for General Microbiology*, American Society for Microbiology, Washington, DC, p. 73.

28   Betz, 1991, *Betz Handbook of Industrial Water Conditioning*, Betz Laboratories, Trevose, PA, p. 279.

29   McNeil, M.B., Jones, J.M., and Little, B.J., 1991, *Corrosion* **47**(9) 674.

30   Crombie, D.J., Moody, G.J., and Thomas, J.D.R., 1980, 'Corrosion of iron by sulfate-reducing bacteria', *Chemistry and Industry* **21** 500.

31   Tatnall, R.E., 1981, 'Fundamentals of bacteria induced corrosion', *Materials Performance* **19**(9) 32.

32   Sanders, P.F., and Hamilton, W.A., 1986, 'Biological and corrosion activities of sulfate-reducing bacteria in industrial plants', in *Biologically Induced Corrosion*, Dexter, S.C., ed., NACE Reference Book No. 8, National Association of Corrosion Engineers, Houston, TX.

33   Bibb, M., and Hartman, K.W., 1984, 'Bacterial corrosion', *Corrosion and Coatings South Africa*, October, 12.

34   McDougal, J., 1966, 'Microbial corrosion of metals', *Anti-Corrosion*, August, 9.

35   Tiller, A.K., 1986, 'A review of the European research effort on microbial corrosion between 1950 and 1984', in *Biologically Induced Corrosion*, Dexter, S.C., NACE Reference Book No. 8, National Association of Corrosion Engineers, Houston, TX.

36   Williams, R.E., Ziomek, E., and Martin, W.G., 1986, 'Surface stimulated increases in hydrogenase production by sulfate reducing bacteria', in *Biologically Induced Corrosion*, Dexter, S.C., ed., National Association of Corrosion Engineers, Houston, TX.

37   Licina, G.J., 1988, *Sourcebook for Microbiologically Influenced Corrosion in Nuclear Power Plants*, EPRI NP-5580, Electric Power Research Institute, Palo Alto, CA, p. 4-4.

38   Licina, G.J., 1988, *Sourcebook of Microbiologically Influenced Corrosion in Nuclear Power Plants*, EPRI NP-5580, Electric Power Research Institute, Palo Alto, CA.

39   Kuhr, C.A.H., and Van der Vulgt, L.S., 1934, *Water*, The Hague, (18), 147, 185.

40   Kashner, D.J., ed., 1978, *Microbial Life in Extreme Environments*, Academic Press, New York.

41   ASM Metals Handbook, 1987, Vol. 13, *Corrosion*, American Society for Metals, Metals Park, OH.

42   Sharpley, J.M., 1961, 'Microbiological Corrosion in Waterfloods', *Corrosion* **17**(8) 386.

43   NACE, TPC 3, 1990, *Microbiologically Influenced Corrosion and Biofouling in Oilfield Equipment*, National Association of Corrosion Engineers, Houston, TX.

44 Licina, G.J., 1988, *Sourcebook for Microbiologically Influenced Corrosion in Nuclear Power Plants*, EPRI NP-5580, Electric Power Research Institute, Palo Alto, CA, p. 2-5.

45 Singha, U., Wolfram, J., and Rodgers, R., 1991, 'Microbially influenced corrosion of stainless steels in nuclear power plants', in *Microbially Influenced Corrosion and Biodeterioration*, Dowling, N.J.E., Mittleman, M.W., Danko, J.C., ed., National Association of Corrosion Engineers, Houston, TX, p. 4-51.

46 Pope, D.H., Duquette, D.J., Johannes, A.H., and Wayner, P.C., 1984, 'Microbiologically influenced corrosion of industrial alloys', *Materials Performance* April, **23**(4), p. 14.

47 Licina, G.J., 1988, *Sourcebook for Microbiologically Influenced Corrosion in Nuclear Power Plants*, EPRI NP-5580, Electric Power Research Institute, Palo Alto, CA, p. 4-4.

48 Kobrin, G., 1986, 'Reflections on microbiologically induced corrosion on stainless steels', in Dexter, S.C., ed., *Biologically Influenced Corrosion*, NACE Reference Book No. 8, National Association of Corrosion Engineers, Houston, TX.

49 Uhlig, H.H., 1948, *Corrosion Handbook*, John Wiley & Sons, New York.

50 Licina, G.J., 1988, *Sourcebook for Microbiologically Influenced Corrosion in Nuclear Power Plants*, EPRI NP-5580, Electric Power Research Institute, Palo Alto, CA, p. 2-7.

51 Licina, G.J., 1988, *Sourcebook for Microbiologically Influenced Corrosion in Nuclear Power Plants*, EPRI NP-5580, Electric Power Research Institute, Palo Alto, CA, p. 2-6.

52 Little, B.J., Wagner, P., and Mansfeld, F., 1992, 'An overview of microbiologically influenced corrosion', *Electrochim. Acta* **37**(12) 2185.

53 Salvarezza, R.C., and Videla, H.A., 1984, *Acta Gientifica Venezalana* **35** 244.

54 ASM Metals Handbook, 1987, Vol. 13, *Corrosion*, ASM International, Metals Park, OH, 44.

55 National Physical Laboratory, 1983, *Microbial Corrosion*, Proceedings of the Conference, The Metals Society, UK.

56 Gaudy, A.F., and Gaudy, E.T., 1980, *Microbiology for Environmental Scientists and Engineers*, McGraw-Hill, New York, p. 353.

57 Gaudy, A.F., and Gaudy, E.T., 1980, *Microbiology for Environmental Scientists and Engineers*, McGraw-Hill, New York, p. 355.

58 Tiller, A.K., 1990, 'Biocorrosion in civil engineering', in *Microbiology in Civil Engineering*, FEMS Symposium No. 59, E. & F.N. Spon, London, p. 26.

59 Tiller, A.K., 1990, 'Biocorrosion in civil engineering', in *Microbiology in Civil Engineering*, FEMS Symposium No. 59, E. & F.N. Spon, London, p. 27.

60 Miller, J.D.A., 1970, *Microbial Aspects of Metallurgy*, American Elsevier, New York.

61 King, R.A., and Miller, J.D.A., 1971, 'Corrosion by the sulfate reducing bacteria', *Nature* **233** 491.

62 Booth, G.H., 1971, 'Corrosion of mild steel by sulfate reducing bacteria: an alternative mechanism', *Brit. Corr. J.* **3** 243.

63 Purkiss, B.E., 1970, 'Corrosion in industrial situations by mixed microbial flora', in *Microbial Aspects of Metallurgy*, American Elsevier, New York.

64 Calderon, G.H., Stratfeld, E.E., and Coleman, C.B., 1968, 'Metal-organic acid corrosion and some mechanisms associated with these processes', in *Biodeterioration of Materials*, American Elsevier, New York.

65 Little, B.J., Wagner, P.A., and Duquette, D., 1987, 'Microbiologically induced cathodic depolarization', *Corrosion/87*, National Association of Corrosion Engineers, Houston, TX.

66 Edyvean, R.G.J., 1990, 'Fouling and corrosion in water filtration and transportation systems', in *Microbiology in Civil Engineering*, FEMS Symposium No. 59, E. & F. N. Spon, London, p. 62.

67 Sharpley, J.M., 1961, 'The occurrence of *Gallionella* in salt water', *App. Microbio.* **9** 380.

68 Carlson, V., Bennett, E.O., and Rowe, J.A., 1961, 'Microbial flora in a number of oilfield water injection systems', *J. Soc. Pet. Engineers* **1** 72.

69 Myers, G.E., and Slabyib, B.M., 1962, 'The microbiological quality of injection waters used in Alberta oil-fields', *Producers Monthly* **26** 12.

70 Geesey, G.G., personal communication.

71 Gerchakov, S.M., and Sallman B., 1976, 'Biofouling and effects of organic compounds and microorganisms on corrosion processes', University of Miami School of Medicine, Miami, FL.

72 Miller, J.D.A., 1970, *Microbial Aspects of Metallurgy*, American Elsevier, New York, p. 88.

73 Mara, D.D., and Williams, J.D.A., 1971, 'Corrosion of mild steel by nitrate-reducing bacteria', *Chem. Ind.* **21** 566.

74 Brock, T.D., Smith, D.W., and Madigan, M.T., 1984, *Biology of Microorganisms*, 4th ed., Prentice Hall, Englewood Cliffs, NJ, p. 817.

75 NACE, 1984, *Corrosion Basics, An Introduction*, National Association of Corrosion Engineers, Houston, TX, p. 12.

76 Costerton, J.W., and Boivin J., 1991, 'The role of biofilms in microbial corrosion', in *Microbially Influenced Corrosion and Biodeterioration*, Dowling, N.J.E., Mittleman, M.W., Danko, J.C., eds., National Association of Corrosion Engineers, Houston, TX, pp. 5–86.

77 Characklis, W.G., and Cooksey, K.E., 1983, 'Biofilms and microbial fouling', in *Advances in Applied Microbiology*, Laskin, A.I., ed., Academic Press, London, p. 93.

78 Niehof, R.A., and Loeb, G., 1973, 'Molecular fouling of surfaces in seawater', in *Proceedings of the III International Congress on Marine Corrosion and Fouling*, Acker, R.F., Brown, B.F., DePalma, J.R., and Iverson, W.P., eds., Northwestern University Press, Illinois, p. 710.

79 Loeb, G. and Niehof, R.A., 1977, 'Adsorption of an organic film at the platnium–seawater interface, *J. Mar. Res.* **35** 283.

80 Hamilton, W.A., 1985, 'Sulfate reducing bacteria and anaerobic corrosion', *Ann. Rev. Microbiol.* **39** 195.

81 King, D.A., and Geary, D., 1985, 'Controlling the internal corrosion of subsea pipelines', in *Advances in Offshore Oil and Gas Pipeline Technology*, Gulf Publishing.

82 ASM Metals Handbook, 1987, Vol. 13, *Corrosion*, ASM International, Metals Park, OH, p. 13.

83 NACE, 1984, *Corrosion Basics, An Introduction*, National Association of Corrosion Engineers, Houston, TX, p. 32.

84 ASM Metals Handbook, 1987, Vol. 13, *Corrosion*, ASM International, Metals Park, OH, p. 42.

85 NACE, 1984, *Corrosion Basics, An Introduction*, National Association of Corrosion Engineers, Houston, TX, p. 31.

86 Little, B., Wagner, P., and Duquette, D., 1988, *Corrosion* **44**(5) 270.

87 Fleming, H.C., and Geesey, G.G., 1991, *Biofouling and Biocorrosion in Industrial Water Systems*, Springer-Verlag, New York, p. 158.

88 NACE, 1984, *Corrosion Basics, An Introduction*, National Association of Corrosion Engineers, Houston, TX.

89 Licina, G.J., 1988, *Sourcebook for Microbiologically Influenced Corrosion in Nuclear Power Plants*, EPRI NP-5580, Electric Power Research Institute, Palo Alto, CA, p. 2-2.

90 Borenstein, S.W., 1988, 'Microbiologically influenced corrosion failures of austenitic stainless steel welds', in *Corrosion/88*, National Association of Corrosion Engineers, Houston, TX.

91 Gilbert, R.J., and Lovelace, D.W., ed., 1975, *Microbial Aspects of the Deterioration of Metals*, Academic Press, London.

92 Tiller, A.K., ed., 1988, 'The impact of microbially induced corrosion on engineering alloys', *Microbial Corrosion*, conference proceedings, The Metals Society, National Physical Laboratory.

93 Weber, G.R., 1983, 'Isolation and testing of metal corroding bacteria', *Materials Performance*, **22**(10), p. 24.

94 Mara, D.D., and Williams, D.J.A., 1972, 'Influence of the microstructure of ferrous metals on the rate of microbial corrosion', *Br. Corr. J.* **7** 139.

95 NACE, 1984, *Corrosion Basics, An Introduction*, National Association of Corrosion Engineers, Houston, TX, p. 27.

96 Gaudy, A.F., and Gaudy, E.T., 1980, *Microbiology for Environmental Scientists and Engineers*, McGraw-Hill, New York, p. 327.

Chapter 3

Metallurgy

Metallurgy is concerned with producing metals and alloys, transforming them into usable products, and improving their performance. It is also involved with the chemical nature of reactions and has been described as a form of solid-state chemistry. These chemical reactions relate to the chemical, physical and mechanical behaviour of metallic materials. According to one common system of classifications, the field of metallurgy can be divided into extractive metallurgy, physical metallurgy and mechanical metallurgy. This chapter describes how a metal's properties and characteristics affect metal corrosion, particularly as influenced by microorganisms.

Extractive metallurgy is the refining of ores, which involves extracting metal content from rocks and minerals that is worth the cost of extracting, and then processing and producing a concentrated form of this material for subsequent use by others, such as by a steel mill.

Physical metallurgy studies the effects of treatments and composition on the structure and properties of metals. For example, a physical metallurgist may work at a steel mill and develop a new kind of stainless steel that has higher strength and better corrosion resistance.

Mechanical metallurgy studies metal processing and mechanical behaviour in service. These three fields are sometimes simplified as two types: process metallurgy and physical metallurgy. The study of metallurgy is commonly divided into ferrous (iron-based) or non-ferrous metallurgy, so separating iron-based alloys from all other alloys, such as aluminium.

3.1   Review of ferrous metals

The distinction of what actually constitutes a metal is often unclear. A pure element is made up of only one type of atom and one crystal structure, but may have some impurity atoms. An alloy is any combination of at least two or more chemical elements, of which one must be a metal. A metal is defined as an opaque, lustrous chemical substance that is a good conductor of heat and electricity. About half the elements are classified as metals by their electronic structure.

Any number of the many alloys that contain iron are termed ferrous. Alloys that contain iron as the primary component, and small amounts of carbon as a major alloying element, are referred to as steel.

3.2 Glossary

The following are frequently used steel industry terms. This information is from the ASM Metals Handbook, 9th ed., Vol. 1, *Properties and Selection: Irons and Steels*. There may be some overlap between meanings.

3.2.1 Classification within the steel industry

- **Class**: strength or surface smoothness.
- **Grade**: chemical composition.
- **Type**: deoxidation practice.

3.2.2 Classification by ASTM specifications

- **Class**: strength levels.
- **Designation**: the specific identification of each grade, type or class of steel by a number, letter, symbol, name or suitable combination of identifying terms to a particular steel.
- **Grade**: sometimes denotes strength. For example, ASTM A515-70 denotes material made to the product form (plate) of ASTM standard A515 with a minimum tensile strength of 483 MPa (70 ksi).
- **Quality**: a term used in the steel industry in product descriptions to imply special characteristics that make the mill product suited to specific applications.
- **Type**: chemical composition.

3.2.3 Metallurgical terms

- **Alloy steels**: steels that contain manganese, silicon, chromium, molybdenum, nickel or copper in quantities greater than those listed for the carbon steels. The alloying enhances mechanical properties, fabricating characteristics or other features.
- **Austenite**: face-centred cubic crystal structure. Austenite is also gamma (γ) phase, a high temperature form of iron.
- **Bainite**: a microconstituent in some steels and cast irons that results from the transformation of austenite and consists of finely dispersed alpha ferrite and cementite. It forms at temperatures between those at which pearlite and martensite transformations occur.

- **Carbon steels**: steels that contain less than 1.65% manganese, 0.60% silicon and 0.60% copper with no minimum for other alloying elements, but usually only a small amount.
- **Cast iron**: a generic term for a group of cast ferrous alloys in which the carbon content exceeds the solubility of carbon in austenite in the eutectic temperature. Usually the carbon is greater than 2% in cast irons. Most contain between 3.0 and 4.5 wt % C.
- **Cementite**: iron carbide (Fe_3C).
- **Delta ferrite**: body-centred cubic crystal structure of iron. The delta ferrite, also called delta (δ) phase, forms at high temperatures and is found as a room temperature constituent of many stainless steels.
- **Ductility**: the measure of a material's ability to undergo considerable plastic deformation before fracture.
- **Elastic modulus** (E): the ratio of stress-to-strain when deformation is totally elastic. It is the measure of a material's stiffness.
- **Eutectic reaction**: a reaction wherein, upon cooling, a liquid phase transforms isothermally and reversibly into two new solid phases that are intimately mixed.
- **Eutectoid reaction**: a reaction wherein, upon cooling, a solid phase transforms isothermally and reversibly into two new solid phases that are intimately mixed.
- **Ferrite**: a body-centred cubic crystal structure. Ferrite, also called alpha (α) phase, is the stable low temperature form of iron in most steels.
- **Hypereutectoid steel**: a steel of carbon content greater than the eutectoid concentration. For a plain carbon steel, the hypereutectoid range is between 0.77 and 2.11 wt % C.
- **Hypoeutectoid steel**: a steel of carbon content less than the eutectoid. For a plain carbon steel, the hypoeutectoid range is between 0.022 and 0.77 wt % C.
- **Martensite**: a metastable iron phase, supersaturated in carbon, which is the product of a diffusionless transformation from austenite.
- **Pearlite**: a microconstituent in some steels and cast irons that results from the transformation of austenite and consists of alternating layers or bundles of alpha ferrite and cementite.
- **Phase transformation**: the change in the number and/or character of the phases that constitute the microstructure of an alloy.
- **Spheroidite**: a microconstituent in some steels, consisting of spherically shaped cementite particles in an alpha ferrite matrix.
- **Steel**: an iron-based alloy usually containing manganese, carbon and other alloying elements. Steel is different from 'irons'. There are two groups of irons: cast irons (high carbon) and pure irons (low carbon).
- **Tempered martensite**: the microstructural product resulting from a tempering heat treatment of a martensitic steel. The microstructure

consists of extremely small and uniformly dispersed cementite particles embedded in an alpha ferrite matrix.

- **Tensile strength**: the maximum engineering stress, in tension, that may be sustained without fracture.

3.2.4   Heat treating terms and processes

- **Annealing**: a heat treatment that consists of heating a metal to, and holding it at, a suitable temperature followed by slow cooling. Annealing commonly refers to a heat treatment by which a cold worked material is softened to allow it to recrystallize.
- **Austenitizing**: forming austenite (a high temperature form of iron) by heating a ferrous alloy above its upper critical temperature to within the austenite phase region from the phase diagram.
- **Full anneal**: for ferrous alloys, austenitizing, followed by cooling slowly to room temperature; coarse pearlite is formed during cooling.
- **Hardenability**: the measure of the depth to which a specific ferrous alloy may be hardened by the formation of martensite. Hardenability describes the ability of an alloy to be hardened by the formation of martensite as a result of a given heat treatment.
- **Jominy end-quench test**: a standard for determining hardenability.
- **Normalizing**: for ferrous alloys, austenitizing above the upper critical temperature, then cooling in air. The objective is to enhance toughness by refining the grain size.
- **Process annealing**: an imprecise term meaning improved workability indicating that the metal was heated below the lower critical temperature to allow recrystallization, but not grain growth. Used for fabrication that requires repeated plastic deformation.
- **Quenching**: the rapid cooling of a steel from the austenitizing temperature by immersing the piece in a liquid or gas.
- **Spheroidizing**: for steels, a heat treatment carried out just below the eutectoid, in which the spheroidite microconstituent is produced.
- **Stress relief**: a heat treatment for the removal of residual stresses.
- **Tempering**: a thermal history of steel in which martensite, formed by quenching austenite, is reheated.

3.2.5   Time temperature transformation terms

- **Precipitation hardening**: hardening and strengthening of a metal alloy by extremely small dispersed particles that precipitate from a supersaturated solid solution; sometimes called age hardening.
- **TTT diagram**: the time temperature transformation diagram is a plot of the time necessary to reach a given percentage transformation at a given temperature (TTT diagrams are also called isothermal transformations or IT diagrams). A TTT diagram for steel shows

transformation products resulting from cooling austenite, including the time to the beginning and the ending of the transformation. The TTT diagram of a steel is essentially a map that charts the transformation of austenite as a function of temperature and time, and shows approximately how a particular steel will respond to any mode of slow or rapid cooling from the austenitic state.

3.3   Introduction to steel

Before the Middle Ages, the only ferrous metal known was iron. The term iron means 'stone from the heaven' in several languages. The first iron was from meteorites, which contained large quantities of metallic iron.

Metallurgists theorize that these meteorites were somehow, possibly by chance, placed around or in primitive wood fires. After the fires cooled, the ashes contained a small amount of iron sponge from the meteorites. This sponge was the iron that remained after the oxygen was driven off the meteorites by the fire's heat, a process known in chemical terms as reduction of iron oxide. The reduction took place because of the carbon in the ashes, which reduced the melting point of the iron so it would melt in the fire.

These pieces of sponge iron could be collected and hammered into shapes, which were relatively strong and ductile. This technique was duly modified into simple furnaces; our ancestors discovered raw iron ore and made it into usable tools. By 1500 BC, the making and shaping of iron for tools was well established in Egypt.

For the next 3000 years, not much changed, metallurgically speaking. Labourers reduced iron oxide into iron sponge and pounded it into a product, producing wrought iron.

In the Middle Ages, the blast furnace was invented. It forced hot air into the iron ore, thereby increasing the rate at which reduction of iron oxide happened, as well as slightly changing the chemistry of the reaction. Use of blast furnaces produced pig iron, which was hard, brittle and difficult to shape by hammering. It was suitable primarily for casting.

The differences between wrought iron and pig iron lie primarily in composition. Wrought iron has about 0.005% or less carbon, while pig iron has 3% or more carbon. When the early metal workers were able to modify their processing to get between 0.1 and about 2% carbon, they discovered steel, which occurred only once furnaces were capable of high enough temperatures to melt the low-carbon iron alloy.

3.3.1   Carbon steel

Steel is an iron-based alloy, different from irons, that usually contains manganese, carbon and other alloying elements. The categories of steel

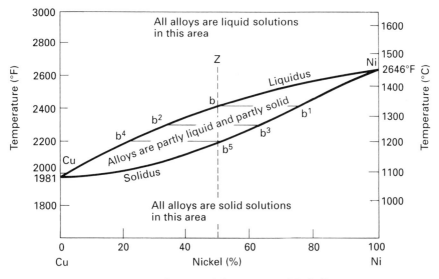

3.1   Constitutional diagram of the copper–nickel alloys.

originate around the chemical compositions, primarily carbon content. Carbon steels contain less than 1.65% manganese, 0.60% silicon and 0.60% copper, with no minimum for other alloying elements, but usually containing only a small amount.

A common method of describing events in a two-component system is via phase diagrams, which plot temperature versus composition. A simple phase diagram is shown in Fig. 3.1, which shows the relationship of temperature versus composition. This particular phase diagram is a copper–nickel alloy, featuring complete solubility of the components in the solid and the liquid states. The upper line is the liquidus, and the alloy is completely liquid above this line. The two end points are the melting points of the pure elements, copper and nickel. The lower line is the solidus, since all compositions below this line are in the solid state. The solidus line shows the temperature and compositions where freezing (and solidification) occurs.

Iron–iron carbide phase diagram

The iron–iron carbide phase diagram is a much more complicated diagram. This diagram details temperatures and compositions for iron as increasing amounts of carbon are dissolved in the iron. Such details tell us how iron changes into steel and cast iron.

Iron is an allotropic metal, existing in more than one type of crystal

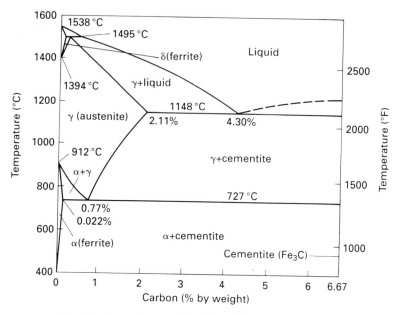

3.2 The iron–iron carbide phase diagram. Because of the importance of steel as an engineering material, this diagram is one of the most important phase diagrams.

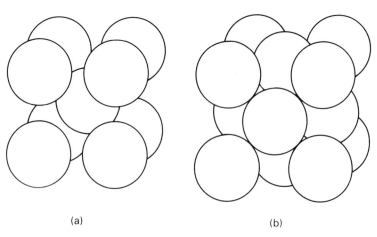

(a)   (b)

3.3 Crystal structure: (a) body-centred cubic unit of structure; (b) face-centred cubic structure (source: Parr, J.G., and Hanson, A.H., 1965, *An Introduction to Stainless Steel*, ASM International, Materials Park, Ohio, p. 19).

structure, depending on the temperature. A primary element influencing iron's crystal structure change is carbon; the temperature at which the changes occur depends on the alloying elements.

Using an iron–iron carbide phase diagram (Fe–Fe$_3$C) (shown in Fig. 3.2), metallurgists can determine the phases present at equilibrium at a specific temperature or composition.[1]

Ferrite (alpha-iron, α) is a room temperature form of iron (see Fig. 3.2). It has a body-centred cubic (BCC) structure (Fig. 3.3(a)) and is stable up to 910°C (1670°F). (The addition of up to about 1.7% carbon changes its properties and makes iron into steel.) The BCC crystal structure allows very little solubility of carbon in this solid state.

Austenite (gamma-iron, γ) is stable between 910 and 1390°C (1670 and 2540°F). It has a face-centred cubic (FCC) structure (Fig. 3.3(b)). The FCC crystal structure allows more solubility of carbon into the structure. Austenite has a denser packing, because it is FCC, than ferrite, which is BCC.

Above 1390°C (2540°F), the structure returns to BCC and is called delta-iron or delta-ferrite, δ. While the solubility of carbon in delta-ferrite is larger than that of alpha-ferrite, it is still relatively low.[1]

To summarize, iron–carbon alloys containing less than 2% carbon are steels. Iron–carbon alloys containing greater than 2% carbon are the cast irons. Additional information about the structure of steels is given in Section 3.4.

The classifications of carbon and alloy steels

Steels are classified by several methods, including by manufacture, use of the alloy in type of product, chemical composition and mechanical properties.

Manufacturing method categories include Bessemer steel, open-hearth steel, electric-furnace steel and crucible steel. Some steel specifications give the particular method of manufacture required to comply with a standard.

Classification by use categories includes machine steel, spring steel, boiler steel, structural steel and tool steel.

Classification by chemical composition is common. Steel specifications by this method are generated by the American Iron and Steel Institute (AISI) and the Society of Automotive Engineers in America (SAE), with a numbering system to indicate the approximate alloying content. Letter prefixes are sometimes included to designate the steel-making process (B = acid Bessemer, C = basic open-hearth and E = basic electric-furnace). The last two numbers indicate the carbon content (AISI 1020 has 0.20% carbon, for example).

Table 3.1. AISI series

10xx	Plain carbon steels
11xx	Carbon steel, high sulphur, low phosphorus
12xx	Carbon steel, high sulphur, high phosphorus
15xx	Manganese to 1.65%
23xx	Nickel 3.50%
31xx	Nickel 1.25%, chromium 0.65 and 0.80%
40xx	Molybdenum 0.20% or 0.25%
41xx	Chromium 0.50%, 0.80% or 0.95%, molybdenum, 0.12%, 0.20% or 0.30%
43xx	Nickel 1.82%, chromium 0.50% or 0.80%, molybdenum 0.25%
46xx	Nickel 0.85% or 1.83%, molybdenum 0.20% or 0.25%
47xx	Nickel 1.05%, chromium 0.45%, molybdenum 0.20% or 0.35%
48xx	Nickel 3.5%, molybdenum 0.25%
51xx	Chromium 1.02%
5xxx	Carbon 1.4%, chromium 1.03% or 1.45%
61xx	Chromium 0.60% or 0.95%, vanadium 0.13% or 0.15%
86xx	Nickel 0.55%, chromium 0.50%, molybdenum 0.20%
87xx	Nickel 0.55%, chromium 0.50%, molybdenum 0.25%
88xx	Nickel 0.55%, chromium 0.50%, molybdenum 0.35%
92xx	Silicon 2.00%

AISI series

Steels specified by AISI (Table 3.1) comply with chemical compositions only, and will not meet strength requirements. Many purchasers mistakenly assume that strength requirements have been met: they must realize that they need to specify such additional requirements as strength.

Carbon steels (1xxx series)

Plain carbon steels are primarily used where strength and other property requirements are not severe. Low alloy steels are carbon steels with added elements to enhance the properties of the steel.

Nickel steels (2xxx series)

Nickel alloys are used for high strength structural steels in the as-rolled condition, and for large forgings that cannot be hardened by heat treatment. Forging is a hot working process, in which metal is made to flow under high compressive stresses, by being smashed into a die cavity during one common method of forging.

Nickel widens the range for successful heat treatment and does not form carbides. The nickel promotes the formation of very fine and tough pearlite at lower carbon contents. Nickel steels provide increased toughness, plasticity and fatigue resistance.

Nickel–chromium steels (3xxx series)

Nickel increases toughness and ductility. These features, combined with the effect of chromium, improve hardenability and wear resistance. Two

alloying elements combined often give results in property improvements greater than the sum of their respective individual features.

Manganese steels (31xx series)

Manganese is a deoxidizer and is added to steel to reduce hot-shortness. When the manganese content exceeds 0.80% it increases the strength and hardness in high carbon steels. Fine-grained manganese steels have excellent toughness and strength.

Steels with greater than 10% manganese remain austenitic after cooling and are known as Hadfield manganese steel. After heat treatment, this steel has excellent toughness and wear resistance, as well as high strength and ductility. Work hardening occurs as the austenite is strain hardened to produce martensite. In early days these were used as jail bar steels: the first cut with a hacksaw work hardened the metal and the next cut broke off all of the teeth of the saw on the hardened metal.

Molybdenum steels (4xxx series)

Molybdenum is a carbide-former and has limited solubility in α and δ iron. It has a strong effect on hardenability and increases high temperature strength and hardness.

Chromium steels (5xxx series)

Chromium forms simple and complex carbides ($(Cr_7C_3$, $Cr_{23}C_6)$ and $(FeCr)_3C$ respectively). The carbides increase hardness and wear resistance. Chromium is soluble up to about 13% in δ iron. It has unlimited solubility in ferrite. It increases strength and toughness of the ferrite, and also improves high temperature properties and corrosion resistance.

The effects of alloying elements for carbon steels

Carbon steels contain sulphur below 0.05%, because the sulphur combines with iron to form FeS, a low melting point alloy that tends to concentrate at grain boundaries. At elevated temperatures, high sulphur steel results in hot-shortness because of melting of the FeS eutectic. These steels literally fall apart when worked (forged, rolled, etc.).

Free-machining steels contain sulphur ranging from 0.08% to 0.35%, which promotes sulphide inclusions that act as chip breakers. The chips break easily which significantly reduces tool wear, in machining operations.

Manganese is present in all commercial steels in the range of 0.03–1.00%. Manganese counteracts sulphur by forming manganese sulphide, which comes off as a slag. Any excess manganese combines with carbon to form the Mn_3C compound, associated with cementite. In addition, manganese acts as a deoxidizer in the melt.

Phosphorus is present below 0.04% because greater amounts reduce the steel's ductility. Additions of 0.07–0.12% phosphorus improve cutting properties.

Silicon contents range from 0.05% to 0.3%. Silicon dissolves in ferrite, which increases the steel's strength while maintaining its ductility. It enhances the effect of carbon. Silicon promotes deoxidation of the steel melt through the formation of SiO_2.

The effects of alloying elements for alloy steels

Alloy steels are used when strength and other property requirements are more demanding than plain carbon steel can handle. They are also used when high temperatures and corrosive environments are factors.

Alloy steels characteristically have properties due to some element other than carbon. The purposes for adding alloying elements include:

- Increased hardenability.
- Improved strength at ambient temperatures.
- Improved mechanical properties at low and high temperatures.
- Improved toughness.
- Increased wear resistance.
- Increased corrosion resistance.
- Improved magnetic properties.

Alloying elements are distributed in the main constituents of steel in two ways:

- Elements that dissolve in ferrite.
- Elements that combine with carbon to form simple or complex carbides.

The effect of alloying elements on ferrite

Nickel, aluminium, silicon, copper and cobalt all dissolve in ferrite and, in so doing, increase the ferrite's strength, usually by a process called solid solution hardening (see Section 3.4). Specifically, the alloying elements change the critical range, eutectoid composition, and the temperature and location of the α and δ fields in the iron–iron carbide phase diagram. These changes affect the heat treating requirements as well as the final properties of alloys.

The effects of alloying elements on carbide

Manganese, chromium, tungsten, molybdenum, vanadium and titanium are carbide-forming elements. Addition of these elements to steel increases room temperature tensile properties, because all the carbides are hard and brittle. Complex carbides are hard to dissolve even at elevated temperatures, so they act as inhibitors to grain growth.

Chromium and vanadium carbides are exceptionally hard. As alloying elements to steel, they promote excellent wear-resistance properties.

The mechanical properties of alloy and carbon steels

Mechanical metallurgy is the science of the behaviour of metals and alloys subjected to applied forces. During manufacturing processes, a part is shaped by applying an external force to it; for example, flat sheet metal is processed into pieces of an auto body by mechanical means. Forming operations may be carried out at various temperatures, at different rates, and with forces in several directions.

Mechanical properties include, for example:

- Hardness.
- Tensile strength.
- Compressive strength.
- Ductility.
- Elasticity.

One approach uses the strength of the material and the theories of elasticity and plasticity, by which a metal is considered a homogeneous material whose mechanical behaviours are described on the basis of only a few material constants.

The theories of strength of materials, elasticity and plasticity lose much of their power, however, when the material's structure becomes important and it can no longer be considered a homogeneous medium. Mechanical properties or mechanical behaviour change as the metal or alloy is processed by welding or heat treating, for instance.

Mechanical properties and phases

An alloy's mechanical properties depend on the properties of the phases, and the way those phases are arranged to make up the structure.

Ferrite is relatively soft with low tensile strength, and cementite (Fe_3C) is hard with very low tensile strength. The combination of these two phases in the form of pearlite produces an alloy with much greater tensile strength than that of either phase, and ductility that is better than the cementite, though not as good as ferrite.

Strength of materials

Strength of materials is the science of the relation between internal forces, deformation and external loads. A member is assumed to be in equilibrium; the equations of static equilibrium are applied to the forces acting on some part of the body, thus obtaining a relationship between the external forces and the internal forces resisting their action.

Metals are composed of crystals, having different properties in different directions. The equations of strength apply, since the crystal grains are so small that materials on a macroscopic scale are statistically homogeneous and are often isotropic.

Elastic and plastic behaviour

All solid materials deform when subjected to an external load. The recovery of the original dimensions of a deformed body when the load is removed is called elastic behaviour. The limiting load beyond which material no longer behaves elastically is the elastic limit, and a body that is permanently deformed is said to have undergone plastic deformation.

For most materials that are loaded below the elastic limit, the deformation is proportional to the load in accordance with Hooke's law:

$$\sigma = E\varepsilon \qquad\qquad [3.1]$$

where σ is stress, the force per unit area, and ε is the strain, the change in length per the original length, E is the elastic modulus, or Young's modulus. This information is determined by performing a tensile test. A stress–strain curve, typical of a standard tensile test, is shown in Fig. 3.4.

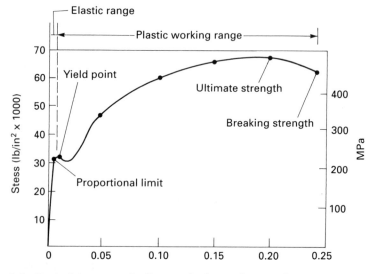

3.4 Typical stress–strain diagram for low carbon steel.

Ductile versus brittle behaviour

Ductility is the measure of a material's ability to deform plastically, i.e. to undergo considerable plastic deformation before fracture. It is expressed as percentage elongation or percentage area reduction. Brittleness is the quality of a material that leads to cracks and failure with very little plastic deformation.

A completely brittle material would fracture almost at the elastic limit. A brittle material such as white cast iron would show some measure of plasticity before fracturing. Adequate ductility is important, since it allows the material to redistribute localized stresses. If localized stresses at notches and other stress concentrations need not be considered, metallurgists can design for static situations on the basis of average stress.

With brittle materials, localized stresses continue to build when there is no local yielding, until a crack forms at one or more points of stress concentration, and spreads rapidly over the section.

Even without a stress concentration, fracture occurs rapidly in a brittle material, since the yield stress and tensile strength are practically identical.

The physical properties of alloy and carbon steels

Physical properties are inherent to a metal or alloy, and will not change with processing or microstructure. Physical properties include:

- Density.
- Melting point.
- Thermal conductivity.
- Electrical and magnetic properties.

3.3.2   Cast iron

Cast irons, generally classified according to their microstructure, are alloys of iron and carbon in the range of 2–6.67%, and silicon of up to 3.5%. Cast iron's dominant feature is how it solidifies, or its solidification morphology, which depends on the carbon content, the alloy and impurity content, the cooling rate during and after freezing, and heat treatment after casting.

Cast irons have low ductility; they cannot be rolled, drawn or worked at room temperature. Cast irons often have lower strength than most steels.

The condition and physical form of the carbon controls the properties of cast iron. Cast irons are classified by structure; ferrite, pearlite, quenched and tempered, and austempered. The equilibrium phase dia-

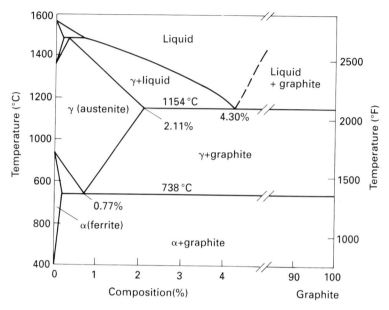

3.5 Phase diagram for the iron–carbon system with graphite, instead of cementite, as the stable phase.

gram for cast iron, shown in Fig. 3.5, is similar to the iron–iron–carbide diagram, but the right boundary is graphite instead of cementite. Also note that cast irons are completely liquid at much lower temperatures than steels. Types of cast iron include:

- White cast iron.
- Grey cast iron.
- Nodular cast iron.
- Malleable cast iron.
- Pearlitic cast iron.

White cast iron

White cast iron contains a large amount of hard, brittle cementite. This microconstituent forms a continuous interdendritic network. White cast iron is hard and wear-resistant, but also brittle and difficult to machine.

Grey cast iron

Grey cast iron is one of the most widely used alloys of iron. The strength of grey cast iron depends on the matrix in which graphite (free carbon) is embedded (Fig. 3.6(a)). The matrix can range from ferrite to pearlite,

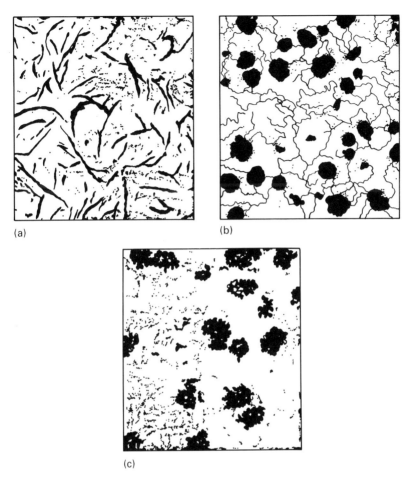

(a)   (b)

(c)

3.6   Microstructure for cast irons, magnification 100×. (1) Ferritic grey iron with graphite flakes. (b) Ferritic nodular iron (ductile iron) with graphite in nodular form. (c) Ferritic malleable iron. This cast iron solidified as white cast iron, with the carbon present as cementite, and was heat treated to graphitize the carbon.

and various combinations of the two phases. Large graphite flakes reduce the strength and ductility, so innoculants are used to promote fine flakes.

ASTM Specification A48 lists seven classes of grey cast irons based on tensile strength. Irons above 275 MPa (40 000 psi) tensile strength are considered high strength irons.

Compressive strength is much more important than tensile strength for many applications. Generally, grey cast iron performs better than steel in compression loading applications.

Nodular cast iron

Nodular iron is also called ductile iron. The graphite is present as tiny spheroids, which interrupt the matrix less than graphite flakes, thus producing a cast iron with higher strength and toughness than grey cast iron (Fig. 3.6(b)).

Nodular iron is produced directly from the melt and does not require heat treatment. Magnesium or cerium is added to the ladle just before casting to cause graphite to form nodules. The matrix can be ferrite, pearlite or austenite. This cast iron has good or better properties than malleable iron and is much less expensive to produce since no energy consuming heat treatment is required.

Malleable cast iron

Although the cementite found in cast iron is considered metastable, heat treatment can transform it into iron and carbon. Malleabilization is performed to convert all of the combined carbon in white iron into irregular nodules of graphite and ferrite (Fig. 3.6(c)). Two annealing stages are performed at about 900–930 °C (1650–1700 °F) for up to 72 hours. The resultant temper carbon is surrounded by a tough ferritic matrix, so that malleable cast iron has higher strength and ductility than grey cast iron.

3.3.3   Stainless steel

Stainless steels do not rust in the same way that most steels rust. The term 'stainless' means a resistance to staining, rusting and pitting.[2] But while it strongly implies that the steel cannot rust, such is not the case.

'Steel' means that the alloy is at least 50% iron. Stainless steel actually includes a large and complex group of different alloys. These alloys, generally a mixture of iron, chromium and nickel, are commonly known for their corrosion resistance.

The corrosion resistance primarily results from a surface chromium oxide film on the metal, known as a passive film, which is stable in an oxidizing environment, but not in a reducing environment. It is also unstable when there is a narrow opening in a metal surface, at a joint between two surfaces, for example, or beneath a solid particle, which would allow for an oxygen concentration cell, resulting in crevice corrosion. The passive film requires oxidizing conditions for stability both to initiate passivity, and to repair subsequent chemical or mechanical damage to the film.[3] Pitting corrosion will probably occur if the passive film is broken.

Chromium is the principal alloying element in making steel into stainless steel. If the chromium content is increased up to 11%, a change

Table 3.2. Stainless steels

2xx	Chromium–nickel–manganese or chromium–nickel–nitrogen
	Nonhardenable, austenitic, nonmagnetic
3xx	Chromium–nickel
	Nonhardenable, austenitic, nonmagnetic
4xx	Chromium
	Hardenable, martensitic, magnetic
4xx	Chromium
	Nonhardenable, ferritic, magnetic
5xx	Chromium
	Low chromium, heat-resisting, magnetic

occurs. As the chromium content increases, an oxide film forms and the corrosion resistance of the steel improves.

Classifications of stainless steels

A three-number system is used to identify stainless steels, as shown in Table 3.2. Stainless steels are divided into five categories: austenitic, ferritic, martensitic, precipitation-hardened and duplex. As previously discussed, the austenitic class of steel is made by adding chromium and nickel and is based on properties resulting from the crystal structure of gamma-iron (austenite).

Austenitic stainless steels

These are the chromium–nickel (Type 3xx) and chromium–nickel–manganese (Type 2xx) stainless steels. Figure 3.7(a) shows the family of austenitic stainless steels.[4] They are austenitic, nonmagnetic and not hardenable by heat treating. They can be hot-worked and cold-worked, but they do readily work-harden into martensite. Austenitic stainless steels are extremely shock-resistant and are difficult to machine. They exhibit the best high temperature strength and scaling resistance of the stainless steels, as well as superior corrosion resistance, when compared to the ferritic and martensitic stainless steels. Type 302 is a basic alloy that has been modified to more than 20 alloys.

Ferritic stainless steels

Ferritic steels similarly result from properties associated with the crystal structure of ferrite. The two terms, 'austenitic' and 'ferritic', refer to either phases, structures or alloy constitutions. The phases have different crystal structures (body-centred versus face-centred) and different interatomic spacings. Figure 3.7(b) shows the family of ferritic stainless steels.

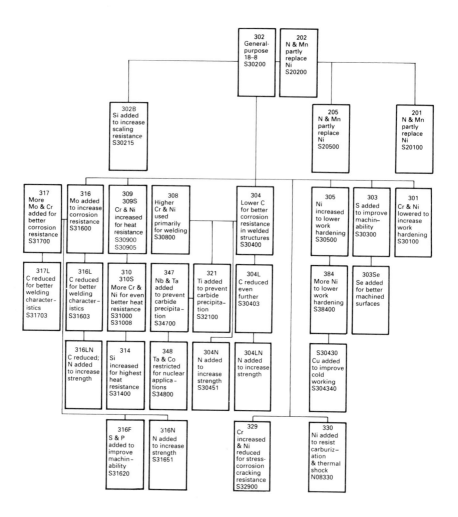

(a)

3.7   Family relationships for (a) standard austenitic stainless steels; (b) standard ferritic stainless steels; (c) standard martensitic stainless steels (1980, 'Properties and selection: stainless steels, tool materials and special purpose metals', in *Metals Handbook*, 9th ed., Vol. 3, ASM International, Materials Park, Ohio, p. 7).

Ferritic stainless steels are straight chromium steels containing between 14% and 27% chromium, such as Types 405 and 430. Since they are so low in carbon content, they are not hardened by heat treatment and can be only moderately hardened by cold working. They have 1.5 times the

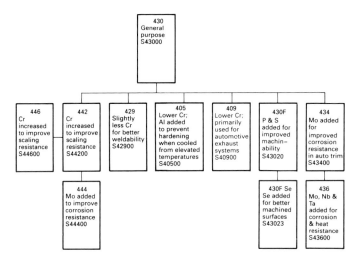

(b)

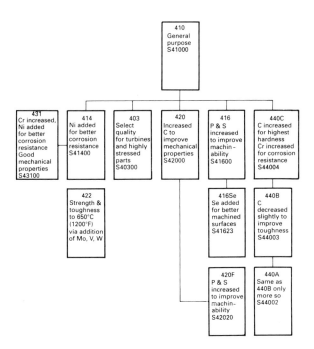

(c)

3.7 *Continued*

strength of carbon steel in the annealed condition, which is also the best condition for corrosion resistance. Prolonged exposure to the temperature range of 400–500 °C (750–950 °F) can cause brittleness and the loss of notch-impact strength.

Martensitic stainless steels

Metallurgists developed martensitic stainless steels to produce a group of alloys that would be corrosion-resistant and hardenable by heat treatment.[5] They did so by adding carbon to the iron chromium system, which produces an alloy that responds to hardening by quenching.[5] These are the straight chromium steels containing 11.5–18% Cr, such as Types 403, 410, 420 and 501. Figure 3.7(c) shows the family of martensitic steels.

Martensitic stainless steels are magnetic. They have good toughness and atmospheric corrosion resistance, and are easily hot-worked. They are most corrosion-resistant when properly heat-treated.

Because of the high alloy content, these materials undergo sluggish transformations and so have high hardenability. Maximum hardness can be achieved by air cooling. Type 416 contains a small amount of sulphur to improve machinability, which also reduces corrosion resistance.

Precipitation-hardening stainless steels

These nonstandard grades were developed during World War II. They are usually supplied in the solution annealed condition and are aged for strength following forming. One commonly used precipitation-hardened stainless steel is 17-4PH, which is solution-treated and air-cooled to allow the austenite-to-martensite transformation. Reheating increases the strength and corrosion resistance. Hardening occurs through the precipitation of small, hard particles (e.g. MO_2 or VC).

Duplex stainless steels

Duplex steels are mixtures of austenite and ferrite. The ferrite phase is desired in the duplex structure to improve weldability, to provide strength and to improve corrosion resistance.[5] Duplex steels are relatively new: they are available as wrought materials (sheet, plate and pipe) and as castings. In addition, weld metal for austenitic stainless steel has a duplex structure.

Alloy constitution

Austenitic stainless steels contain 17–25% chromium, 8–20% nickel and about 2% manganese, along with smaller amounts of carbon, nitrogen, sulphur and phosphorus. An example is Type 304 stainless steel. The addition of 2% molybdenum makes Type 316. Limiting the carbon to

Table 3.3. Typical properties of 300 series austenitic stainless steel

Property	Typical value
Ult. tensile strength	585–655 MPa (85 000–95 000 psi)
Yield strength	240–275 MPa (30 000–40 000 psi)
Elongation in 50 mm (2 in.)	50–60%
Reduction in area	60–70%

0.03% results in an L-grade. These steels are also referred to as 18–8 steels, because they have about 18% chromium and 8% nickel.

Because of their resistance to attack by water, steam, steam condensate, ammonia and oxygen, austenitic stainless steel alloys of Types 304, 304L, 316 and 316L are commonly selected for pipes.[6] These types are also the most weldable of the austenitic stainless steels. The austenitic stainless steels cannot be hardened by heat treatment (like the martensitic) but can be cold-worked (plastically deformed). Austenitic stainless steels are the most commonly used stainless steel alloys because of their corrosion resistance, relatively high strength and resistance to oxidation at high temperatures. Some of the most important properties are listed in Table 3.3.

3.4   The heat treatment of steels

Heat treatment is defined as 'a combination of heating and cooling operations, timed and applied to a metal or alloy in the solid state in a way that will produce desirable properties.'[7] The first step in the heat treatment of steel is heating the material to some temperature at or above the critical range, to form austenite. This treatment then acts to produce phase transformations that influence mechanical properties.

The results of heat treatment depend on:

* Composition.
* Microstructure.
* The amount of cold work prior to heat treatment.
* The rate of heating and cooling.

The heat treatments we will discuss are annealing, quenching, normalizing and tempering. We will also cover the microstructural changes in the iron–carbon system. These structures include:

* Pearlite.
* Spheroidite.
* Bainite.

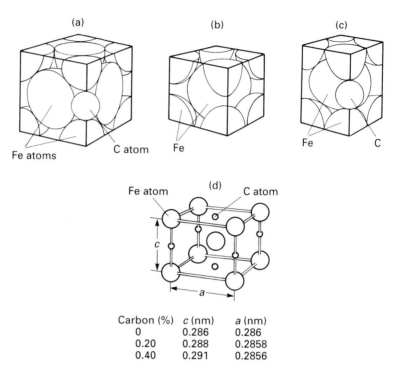

Carbon (%)	c (nm)	a (nm)
0	0.286	0.286
0.20	0.288	0.2858
0.40	0.291	0.2856

3.8 The unit cell for (a) austenite; (b) ferrite; (c) martensite. The effect of percentage of carbon (by weight) on the lattice dimensions for martensite is shown in (d). Note the interstitial position of the carbon atoms and the increase in dimension c with increasing carbon content. Thus the unit cell of martensite is in the shape of a rectangular prism.

- Martensite.
- Tempered martensite.

To review from Section 3.3.1, using the phase diagram:

- Ferrite (α-iron) is a room temperature form of iron (see Fig. 3.8), with a BCC structure and is stable up to 900 °C (1670 °F). The solubility of carbon in alpha ferrite is low.
- Austenite (γ-iron), a high temperature form of iron, is stable between 900 and 1400 °C (1670 and 2540 °F). It has an FCC structure, which allows more solubility of carbon into the structure. Austenite has a denser packing because it is FCC, than ferrite, which is BCC.
- Above 1400 °C the structure is called δ-iron or δ-ferrite. It returns to BCC. While the solubility of carbon in δ-ferrite is larger than that of α-ferrite, because of the expansion of the crystal lattice at the high temperature, it is still relatively low.[1]

3.4.1 *Annealing*

Annealing is a general term that describes heating to and holding at a suitable temperature, and then cooling at a suitable rate. It reduces hardness, improves machinability, facilitates cold-working, produces a desired microstructure, and obtains specific mechanical and physical properties.

Pearlite

Pearlite is a structure formed of layers of alpha ferrite and cementite that often occurs in steel and cast iron.

Cementite is a hard brittle intermetallic compound of iron and carbon, Fe_3C. Pearlite is a microconstituent in some steels and cast irons that results from the transformation of austenite, and consists of bundles of alternating layers of alpha ferrite and cementite, as illustrated in Fig. 3.9.

Isothermal-transformation (I-T) diagrams

Since the phase diagram is of little use for steels that have been cooled under nonequilibrium conditions, I-T diagrams have been developed to predict nonequilibrium structures.

The I-T diagram for a 0.8% carbon eutectoid steel is discussed below.

Above the A_e (austenite end), austenite is stable. The area to the left of the beginning of transformation consists of unstable austenite. The area to the right of the end-of-transformation line is the product to which austenite will transform at constant temperature. The area between the beginning and the end of transformation labelled $A + F = C$ consists of three phases: austenite, ferrite and carbide.[8]

These diagrams are called isothermal-transformation diagrams of time–temperature-transformation diagrams. The diagrams are constructed from data, as shown in Fig. 3.10(a).[8] The percentage of austenite transformed

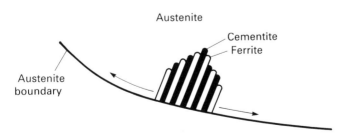

3.9   Schematic picture of the formation and growth of pearlite.

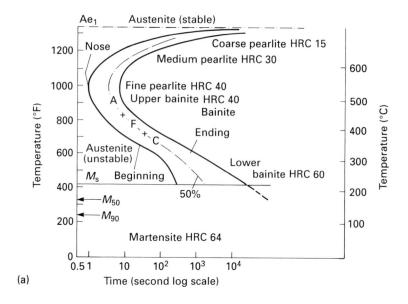

(a)

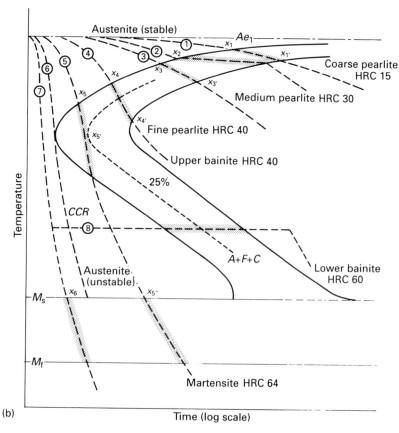

(b)

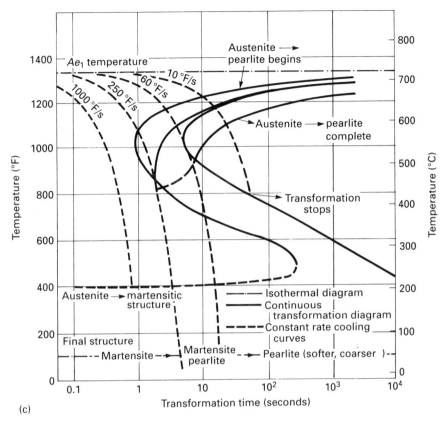

3.10 (a) Isothermal–transformation diagram for 1080 steel; (b) cooling curves superimposed on I–T diagram; (c) continuous cooling–transformation diagram derived from I–T diagram for plain carbon eutectoid steel.

into pearlite is a function of temperature and time: the higher the temperature and/or the longer the time, the more austenite is transformed to pearlite.

The temperature of the start of martensite transformation is known as M_s temperature and at the end as the M_f temperature. The M_s temperature is indicated as a horizontal line, and temperatures for 50% and 90% transformation from austenite to martensite are noted.

The transformation product above the nose region is pearlite. As the transformation temperature decreases, the spacing between the carbide and ferrite layers decreases, and the hardness increases.

Between the nose region of 500 °C (950 °F) and the M_s temperature, an aggregate of ferrite and cementite appears that is called bainite. As the transformation temperature decreases, the bainite structure becomes finer.

Cooling curves

Various cooling curves can be superimposed on the I-T diagram, as shown in Fig. 3.10(b). Curve 1 shows a very slow cooling rate typical of annealing.[9] The material will remain austenitic for a long time; transformation starts when the cooling curve crosses the beginning of transformation at point x_1. Coarse pearlite is formed. Transformation continues until point x'_1, and the pearlite last formed will be slightly finer. Further cooling produces no changes.

Curve 3 is a faster cooling rate typical of normalizing. Transformation starts at x_3 with the formation of coarse pearlite, and is completed at point x'_3, with the formation of medium pearlite.

Curve 5 is an intermediate cooling rate (oil quench); transformation to pearlite at point x_5 occurs in a short time. A change in slope occurs at point x'_5; transformation cannot occur below that point. The microstructure at this point is about 25% fine pearlite surrounded by the existing austenite grains. After the temperature reaches the M_s line, the remaining austenite is transformed into martensite.

Curve 6 is a drastic quench, and transformation in the nose region is avoided. The austenite remains until the M_s line is reached, at which time all of the austenite is transformed to martensite down to the M_f line (martensite finish). A cooling rate that just misses the nose of the curve (curve 7) is the critical cooling rate, and is required to achieve 100% martensite for maximum strength and hardness.

Cooling-transformation (C-T) diagrams

Since the actual heat treatment of steel involves continuous cooling, the C-T diagram (Fig. 3.10(c)) has been derived from the I-T diagram. In general, the nose will be shifted downwards and to the right as a result of faster cooling rates.

Since C-T diagrams are difficult to derive and I-T diagrams normally err conservatively, I-T diagrams are often used to predict structures and to determine critical cooling rates.[10]

Chemical composition and austenite grain size affect the position of the I-T curve. Increasing carbon or certain alloying elements moves the curve to the right, transformation is retarded, and critical cooling rates are slowed down, making martensite easier to obtain. Adding carbide formers can shift the curve to the left.

Spheroidite

Spheroidite is a microconstituent in some steels, consisting of spherically shaped cementite particles in an alpha ferrite matrix. If pearlite is heated to just below the eutectoid temperature and held for a rather long time (such as a day), the cementite transforms into spherical shapes. These are called spheroidites. They create less stress concentration because of their shape. This structure has higher toughness and lower hardness than the pearlite structure and it can be more easily cold-worked.

Bainite

Bainite is a microconstituent in some steels and cast irons that results from the transformation of austenite and consists of alpha ferrite and cementite. It forms at temperatures between those at which pearlite and martensite transformations occur. It is generally stronger and more ductile than pearlitic steels at the same hardness level.[11]

Martensite

Martensite (Fig. 3.8(c)) is a metastable iron phase, supersaturated in carbon, which is the product of a diffusionless transformation from austenite. When austenite is quenched in water, it transforms from an FCC structure to a tetragonal BCC, and is called martensite. It is of limited use because it is extremely hard and brittle.

Tempered martensite

Tempering – heating and improving toughness and reducing hardness – improves martensite's properties. Tempered martensite is the microstructural product resulting from a tempering heat treatment of a martensitic steel. The microstructure consists of extremely small and uniformly dispersed cementite particles embedded in a reduced carbon martensite matrix.

Full annealing

This process consists of heating the steel to the proper temperature and then cooling it slowly through the transformation range in the furnace. Annealing produces a refined grain to induce softness, improves electrical and magnetic properties, and sometimes improves machinability.

Annealing is a slow process that approaches equilibrium conditions and comes closest to following the equilibrium phase diagram.

Spheroidizing annealing

In hypereutectoid steels, the cementite network is hard and brittle, and must be broken by the cutting tool during machining. Spheroidize annealing produces a spheroidal or globular form of carbide and improves machinability. All spheroidizing treatments involve long times at elevated temperatures.

Stress relief annealing

This process is used to remove residual stresses due to heavy machining or other cold-working processes. It is usually carried out at temperatures below 530–650 °C (1000–1200 °F). The term stress relief should not be used interchangeably with annealing.

Normalizing

Normalizing (Fig. 3.11) is carried out at about 38 °C (100 °F) above the upper critical temperature (A_3 line), followed by cooling in still air. Normalizing produces a harder and stronger steel, improves machinability, modifies and refines cast dendritic structures, and refines the grain size for

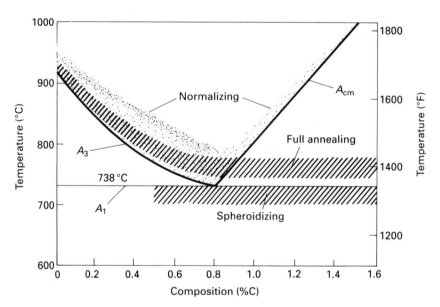

3.11 Heat-treating temperature range for plain-carbon steels, as indicated on the iron–iron carbide phase diagram (source: ASM International).

improved response to later heat treatment operations. Since cooling is not performed under equilibrium conditions, deviations from the structures predicted by the phase diagram do occur.

Hardening by martensite transformation

Under slow or moderate cooling rates, carbon atoms have time to diffuse out of the FCC austenite structure; the iron atoms can then rearrange themselves into the BCC lattice. This transformation takes place by nucleation and growth, and is time-dependent.

Faster cooling rates do not allow sufficient time for the carbon to diffuse out of solution, and the structure cannot transform to BCC with the carbon atoms trapped in solution. The resultant structure, martensite, is a supersaturated solid solution of carbon trapped in a body-centred tetragonal structure. This is a highly distorted crystal structure that results in high hardness and strength.

Martensite atoms are less densely packed than austenite atoms, so a volumetric expansion occurs during transformation. As a result, high localized stresses produce distortion in the matrix. The transformation is diffusionless, and small volumes of austenite suddenly change crystal structure by shearing actions.

Martensite transformation cannot be suppressed or the M_s temperature changed by changing the cooling rate. M_s is a function of chemical composition only.

Martensite is never in a state of equilibrium, although it can persist indefinitely at or near room temperature. Martensite decomposes into ferrite and cementite when heated.

The influence of quenching medium, specimen size and geometry

Other factors influence the extent to which martensite will form in the heat treatment of steel. These factors include quenching medium, specimen size and part geometry. The relationships between cooling rate and specimen size and geometry for a specific quenching medium are frequently expressed on empirical charts. These may be used with hardenability data to generate cross-sectional hardness profiles.

3.4.2 *Quenching*

Cooling proceeds through three separate stages during a quenching operation:

- Vapour-blanket cooling describes the first cooling stage, when the quenching medium is vaporized on the metal surface and cooling is relatively slow.

- Vapour-transport cooling starts when the metal has cooled down enough so that the vapour film is no longer stable and wetting of the metal surface occurs. This is the fastest stage of cooling.
- Liquid cooling starts when the surface temperature of the metal drops just below the boiling point of the liquid, so that vapour is no longer formed. This is the slowest stage of cooling.

Quenching media

The severity of quench indicates the rate of cooling. In order of decreasing severity, the quenches are:

- Brine (10% salt solution).
- Water.
- Fused or liquid salts.
- Soluble oil and water solutions.
- Oil.
- Air.

3.4.3   Tempering

Steel in the as-quenched martensitic condition is too brittle for most applications. High residual stresses are induced as a result of the martensite transformation, and therefore, hardening is nearly always followed by tempering.

Tempering involves heating the steel to some temperature below the lower critical temperature and thus relieving the residual stresses, and improving the ductility and toughness of the steel. There is some sacrifice of hardness and strength. Hardness decreases and toughness increases as the tempering temperature is increased. However, toughness drops when the tempering is performed in the 200–400 °C (400–800 °F) range. Residual stresses are largely relieved during tempering in the temperature range between 200 and 500 °C (400 and 900 °F).

Certain alloy steels exhibit temper brittleness when tempered in the 530–670 °C (1000–1250 °F) range followed by slow cooling. Molybdenum has a retarding effect on temper embrittlement.

As discussed above, martensite is a supersaturated solid solution of carbon trapped in a body-centred tetragonal (BCT) structure and is metastable. As energy is applied during tempering, the carbon is precipitated as carbide and the BCT iron becomes closer to the BCC structure. As the tempering temperature is raised, diffusion and coalescence of the carbides occur and the final product is called tempered martensite.

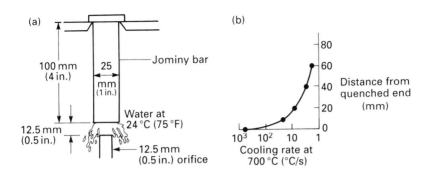

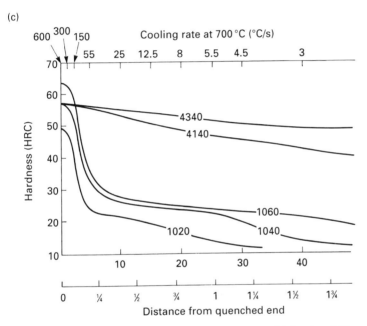

3.12 (a) End-quench test; (b) cooling rate; (c) hardenability curves for five different steels, as obtained from the end-quench test. Small variations in composition can change the shape of these curves. Each curve is actually a band, and its determination is important in heat treatment of metals for better control of properties (source: Lawrence H. VanVlack, *Elements of Materials Science and Engineering*, 5/e (Figures 11-8.2a; 11-8.3; 11-8.4), © 1985 by Addison-Wesley Publishing Company, Inc. Reproduced by permission).

3.4.4   The hardenability of ferrous alloys

Hardenability, the capability of an alloy to be hardened by heat treatment, is the measure of the depth of hardness obtainable by heating and subsequent quenching. The term hardenability should not be confused with hardness, which is an alloy's resistance to indentation.

Hardenability testing

Hardenability is related to the depth of formation of martensite, and can be predicted by the Jominy test. A 25 mm round specimen, 100 mm long (1 × 4 inches) is heated uniformly to the proper austenitizing temperature (to form 100% austenite) and then quenched by a controlled water spray.

Figure 3.12 shows a plot of the hardness versus distance from the quenched end. Since each spot on the test piece represents a certain cooling rate, and since the thermal conductivity of all steels is assumed to be the same, the hardness at various distances can be used to compare the hardenability of a range of compositions.

3.5   Welding

Welding is used to fabricate metal components. The welding process varies with the metallurgical characteristics of the steel or metal. Welding metallurgy studies the behaviour of metal during welding and the effects of welding on the properties of metals.[12]

Welding itself is 'a joining process that produces a coalescence of materials by heating them to the welding temperature, with or without the application of pressure or by the application of pressure alone, and with or without the use of filler metal.'[12] Welding has been practised since earliest recorded history, and was used primarily for fabricating jewellery, tools, weapons and cooking utensils. Modern welding practices were discovered in about 1881, when Moissan used a carbon arc for melting metals.[13] In 1887, Bernandos used this technique of the arc for welding. Soon after, Slavianoff began experimenting with consumable electrodes. Le Chatelier introduced the oxyacetylene blowpipe in 1895. The techniques developed rather slowly until the 1930s, when industrial efforts began to resolve the problems associated with welding. With the advent of World War II, welding was used extensively for tanks, aircraft and shipbuilding. Extensive research and development in welding began during World War II and continues to the present day.

3.5.1   Fundamentals

Welding is often referred to as the art and science of joining metals by use of adhesive and cohesive attractive forces between metals. There are

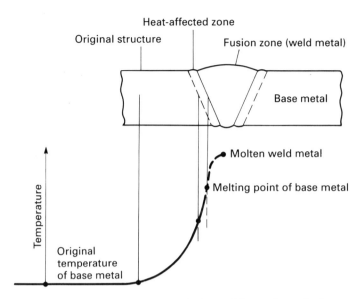

3.13 Characteristics of a typical fusion weld zone in oxyfuel gas and arc welding.

three joining processes, welding, brazing and soldering, which produce metallurgical bonds. Brazing and soldering are similar joining processes. Brazing is defined as the joining of metals with a filler metal melting at a temperature above 450 °C (840 °F) but below the melting point of the metal being joined. Soldering is done below 450 °C.

There are two main categories of welding, fusion and pressure and over 40 commonly used welding processes.[14] Common ones include arc, gas and resistance welding.

Weldability, the capacity of metals to be joined satisfactorily, can be determined via three features:

- The metallurgical compatibility of the materials being joined with a specific process.
- The ability to produce mechanical soundness.
- Serviceability under special requirements.[14]

A typical fusion weld joint is shown in Fig. 3.13 and comprises:

- Base metals (the metal to be welded).
- Heat-affected zone (HAZ).
- Weld metal (the metal that melted during welding).

The metallurgy and the properties of the welded joint depend strongly on the metals joined, the welding process, filler metals used and the welding process variables. A joint produced without filler metals is termed an autogenous weld.

Most welding processes have similar features. All welding processes provide three functions:[15]

- A source of heat that raises the parts to the temperature at which they are welded.
- A source of protection or shielding for the weld to prevent undue contamination from various sources during welding.
- A source of chemical elements that can change, beneficially or detrimentally, the nature of the metal being welded.

3.5.2 Characteristics of weld solidification

Figure 3.14 shows how a typical weld is made. There are cetain generalities in the behaviour of the materials being joined during welding, and in the physical events that take place during welding:[15]

- The base metal is partially melted, and the regions near the weld are heated to high temperatures lower than the melting point.
- A miniature casting is the weld metal, which upon solidification becomes an ingot that should possess certain desirable characteristics.
- Stresses and strains result from the conditions encountered during welding (related to the temperature changes and solidification).
- Some chemical reactions can occur, such as alloying or oxidation.

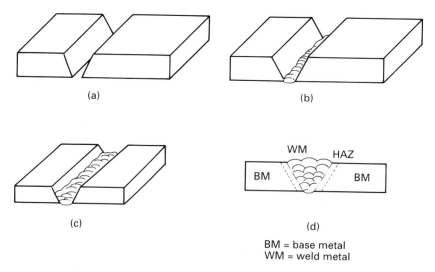

BM = base metal
WM = weld metal

3.14 Schematic of the way a weld is made: (a) prepared weld joint; (b) root pass of weld; (c) fill passes of weld; (d) completed weld.

- The temperature changes are extremely rapid and usually localized.

According to the American Welding Society, welded joints consist of two basic areas: the HAZ, which includes the volume of unmelted base metal immediately adjacent to the weld metal nugget, and the cast weld metal. The properties of the HAZ are determined by the composition of the base metal, and by the heating and cooling cycles resulting from welding or post-heat treatment. The properties of the cast weld metal are determined by chemical composition, dissolved gases and thermal cycles, assuming that this metal is sound and free from cracks or porosity.

Fusion welding is similar to casting. During fusion welding, a number of steps happen rapidly. A volume of molten metal is formed (a casting) within the confines of a solid base metal (or mould). The base metal may be preheated to slow down the cooling rate of the weld joint, just as casting moulds are preheated to slow down cooling during the casting operation. Upon solidification, the weld deposit or nugget (ingot) may be placed in service in the as-welded condition (as-cast) or may be peened, worked (wrought) or heat treated to relieve residual stresses, improve the metallurgical conditions, and provide better properties than the as-welded material.

Let us discuss the weld melt during fusion welding of 0.20% carbon steel. As the flame or arc is directed at the steel, the temperature rises until the microstructure is a mixture of ferrite and fine pearlite. The heat absorbed from the heat source spreads beyond the spot directly underneath. The temperature rises until about 500 °C (950 °F), and, at that point, the grains that were distorted from previous working begin to recrystallize. Inside the 850 °C (1560 °F) curve, the steel is in the gamma (γ) form, and between 850 °C and 725 °C (1335 °F) the steel is a mixture of alpha (ferrite) and gamma (austenite) crystals. Above 900 °C (1650 °F), the austenite crystals grow, as many small austenite crystals coalesce to form larger austenite crystals. At 1490 °C (2714 °F), melting begins; the steel around the weld metal is semisolid, or mushy, since it is a mixture of solid crystals and liquid that continues to 1520 °C (2768 °F), when the last of the crystals melt.

As freezing begins at 1520 °C, crystals of austenite begin to appear and, as cooling continues, the proportion of austenite crystals increases and the melt becomes semisolid (mushy). The tiny crystals replace the liquid randomly (Fig. 3.15). The crystals grow in preferred directions until they come into contact with another crystal growing from a different direction, forming the polycrystalline metal. The first crystals to solidify are usually columnar, while the later crystals are equiaxial. Between 1490 °C and 850 °C, the austenite at the austenite grain boundaries is unchanged. Just above 725 °C, 25% of the steel is austenite and 75% ferrite. Cooling then continues until, at 725 °C, the austenite transforms to pearlite.

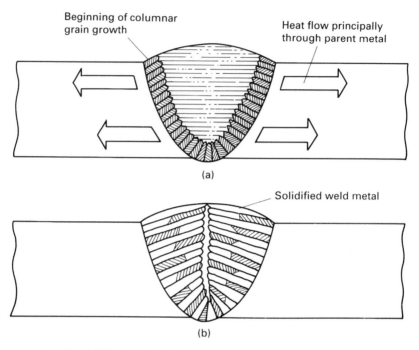

Beginning of columnar grain growth

Heat flow principally through parent metal

(a)

Solidified weld metal

(b)

3.15 Solidification in single pass arc weld: (a) nucleation and columnar grain growth; (b) solidified weld metal.

All of the steel plate that reached a temperature above 1100 °C (2000 °F) will have the structure described above, because the austenite transformations have occurred. The portion of the steel that reached the temperature of about 850–1100 °C would have a more ductile structure. The faster cooling rates in this area would have caused very small austenite grains to form, and the ferrite grains would be distributed throughout within the pearlite grains.

In the steel that reaches 725–850 °C the pearlite grains would transform to austenite and some of the ferrite would be absorbed into the austenite grains. During cooling, small ferrite grains form and, at 725 °C, the remaining austenite transforms back to pearlite. The final structure consists of coarse ferrite grains that had never dissolved, and much finer clusters of ferrite and pearlite grains, where previously only pearlite had existed.

Weld microstructure

Figure 3.16(a) is a schematic representation of the crystallization process. During welding when an arc is directed at a metal surface, the heat is absorbed and spreads as shown in Fig. 3.16(b) and described in the above

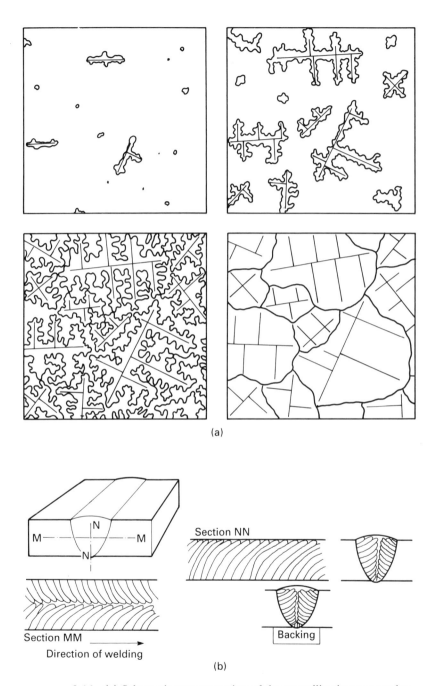

3.16 (a) Schematic representation of the crystallization process by nucleation and dendritic growth. (b) Columnar crystal growth. (c) Austenite; (d) ferrite and (e) pearlite microstructures.

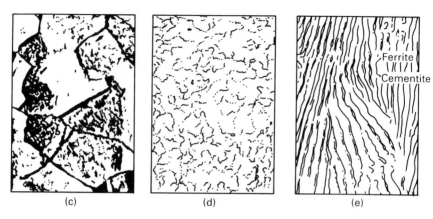

(c)   (d)   (e)

3.16 *Continued*

description on the characteristics of weld solidification. Columnar crystal growth occurs. Figure 3.16(c), (d) and (e) show the microstructure for the three stages in the cooling of austenite described above.

The particular microstructure of a metal determines its properties.[16] The microstructure is primarily determined by heat treatment, alloy composition and fabrication history. Welding heating and cooling rates are extremely important in determining the metal microstructure at or near the weld.

Temperature changes in welding

Welding, particularly fusion welding, produces a complex temperature environment, which in turn produces a wide variety of heat treatments in a weldment.[17] In terms of metal properties, the harmful effects of applying heat for welding outnumber the benefits. It is possible to minimize the undesirable effects of applying heat by controlling the heating and cooling cycles. The following are important considerations for evaluating the thermal effects of welding:

- The rate of heating.
- The length of time at temperature.
- The maximum temperature.
- The rate of cooling.
- The cooling end-point.

Metals are good conductors of heat. Welding results in very rapid cooling in comparison to casting or heat treating, so metals exhibit a metallurgy unique to welding. The atoms pass the heat along to neighbouring atoms

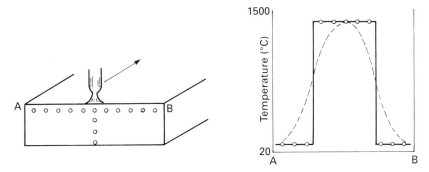

3.17   Temperature distribution in a workpiece during welding.

very readily. Figure 3.17 shows the temperature distribution in a work-piece during welding. Note that the temperature distribution is actually occurring over a cross-section that is constantly changing as the heat source moves along a weld.

Residual stresses

Residual stress is the internal stress that remains in a member of a weldment after a joining operation, such as welding, brazing or soldering. Residual stresses are generated by localized partial yielding during the thermal cycle of welding, and by the hindered contraction of these areas during cooling.

Stresses in the weldment develop primarily because of the local rise and fall in temperature at any point along the joint. As the heat source first advances towards a point and then passes it, stresses and changes in the microstructure develop.

There are two forms of dimensional change. Expansion is a thermal dimensional change, because of simple lattice expansion of the heated atoms. The degree to which it occurs as a function of temperature is referred to as a coefficient of thermal expansion. Volume expansion occurs as parts are heated. Problems occur because the welded parts are restrained. The behaviour of steel, for example, as it attempts to expand or contract in all directions under conditions of restraint, must be considered for a successful welding operation.

The second form of dimensional change may occur as a result of phase transformations. Metals are an array of atoms conforming to a crystal structure. The structure may exist in more than one array at different temperatures. If a phase transformation occurs, a volume change occurs as well, because the atoms are packed differently and different densities will occur.[18]

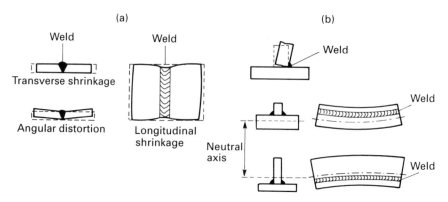

3.18   Distortion of parts after welding: (a) butt joints and (b) fillet welds. Distortion is caused by differential thermal expansion and contraction of different parts of the welded assembly. Warping can be reduced or eliminated by proper fixturing of parts prior to welding.

Residual stresses can cause:[19]

- Distortion, warping and buckling.
- Stress corrosion cracking.
- Further distortion, if a portion of the welded structure is removed (for example by machining).

Residual stresses are usually three-dimensional (Fig. 3.18) and difficult to analyse.

3.5.3   Welding processes

The primary welding processes are oxyfuel gas, arc and resistance welding. We will limit our discussion to arc welding.

In arc welding, the heat is produced by an electrical discharge. The welding arc is discharged through an ionized gas, with the arc produced between the tip of the electrode and the workpiece to be welded. Either AC or DC power supplies can be used, as well as either a consumable or nonconsumable electrode, which is usually a rod or wire.

SMAW

Shielded metal-arc welding (SMAW), also referred to as 'stick welding', is one of the most commonly used and most versatile industrial processes. The electric arc is generated by touching the tip of a flux-coated electrode against the workpiece and then quickly pulling it away slightly to keep the arc. The electrodes are thin long sticks, hence the name.[20] The electrode

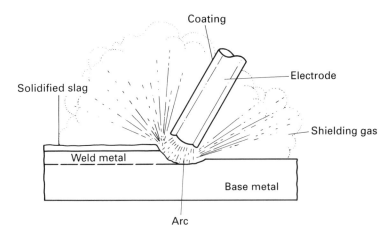

3.19   Schematic illustration of the shielded metal-arc welding process. About 50% of all large-scale industrial welding operations use this process.

becomes the filler metal as it melts off from the heat of the arc. The liquid weld puddle is shielded by both the gaseous products of the flux and a slag formed from the same flux.[21] The weld forms after the molten metal (the base metal (or work-piece), the electrode metal and substances from the coating) is solidified in the weld area.[22] See Fig. 3.19.

GMAW

Gas metal-arc welding (GMAW) was developed in the 1950s and is also called 'MIG' because it used to be referred to as metal inert gas welding. The process is rapid, versatile and economic, as is SMAW, but the equipment to use this process is much more costly than for SMAW. The productivity is about double that for SMAW.

In GMAW, the weld area is shielded by an inert gas such as argon. The consumable bare wire is fed through a nozzle into the weld arc.[23] See Fig. 3.20. The electrodes are especially designed with deoxidizers (silicon, manganese, etc.), to reduce or prevent oxidation of the molten weld puddle.

FCAW

Flux-cored arc welding (FCAW) is similar to GMAW, but the electrode is filled with flux. The flux powder is more flexible than the brittle coating on stick electrodes so the electrode can be coiled onto a spool.[24] See Fig.

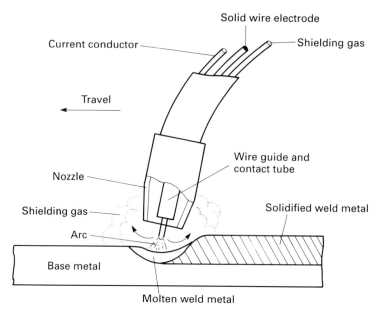

3.20   Gas metal-arc welding process.

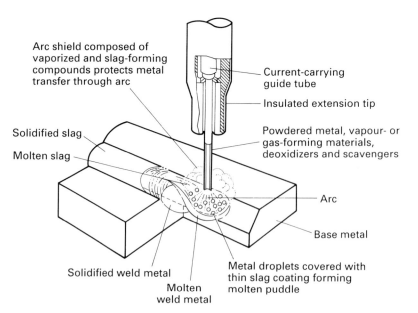

3.21   Schematic illustration of the flux-cored arc welding process.

3.21. The process provides the versatility of SMAW with the continuous feeding features of GMAW.

GTAW

Gas tungsten-arc welding (GTAW) is also known as 'TIG' welding because it was derived from tungsten inert gas welding. The other three processes discussed above used a consumable electrode. The GTAW is a non-consumable electrode process. As one pole of the arc, the tungsten electrode generated the heat required for welding. This process also uses a shielding gas such as argon.[25] See Fig. 3.22.

3.5.4 Heat input

From a practical point of view, the thermal behaviour (the way the temperatures vary with time in the weld and HAZ) is the most important physical characteristic of a fusion weld.[26] The act of welding produces a complex temperature profile, which results in a range of heat treatments to the area beside the weld and rapid cooling. All metals respond to heat treatment by changing their properties in varying amounts. Because of the nature of the welding process with its concentrated heat source, the temperatures change very rapidly and significantly.

The term heat input, also known as energy input, describes the amount of heat energy used for each inch of weld made during a welding operation.[27] The heat input can be determined by the following formula:

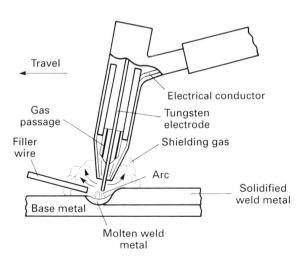

3.22   The tungsten-arc welding process.

$$H = (E \times I \times 60)/S \times 1000 \qquad [3.2]$$

where H is in kilojoules per inch, E is volts, I is amperes, and S is the welding travel speed in inches per minute.

Calculating heat input provides a measure of the heat necessary to make a weld. In addition, it provides a measure of how the heat changes as the parameters (current, voltage and speed) change. It also partially indicates the heating and cooling conditions of a weld.

3.6   The welding of stainless steels

Austenitic stainless steels are the most weldable, relative to other stainless steels. Strength requirements, as well as the factors of corrosion and oxidation resistance, often necessitate that the weld metal composition of stainless steels match fairly closely the base metal composition.[28] To avoid problems when welding, one must understand the properties and characteristics of the metals.

Compared with ferritic steels, austenitic steels have a higher coefficient of expansion, higher electrical resistivity and lower thermal conductivity. For example, the lower thermal conductivity of stainless steel results in the heat of welding concentrating around the weld. Therefore, these alloys often require the use of a low heat input welding process.

3.6.1   Weld microstructure of stainless steels

The overall arrangement of grains, boundaries and phases present in a metal alloy is termed the microstructure, and is largely responsible for the properties of the alloy.[28] Weld microstructure depends on the chemical composition, the solidification mode and the cooling rate.[29] Depending on how the austenitic stainless steel weld is made and the composition, the microstructure is either fully austenitic or duplex (austenite plus delta ferrite).

Phase diagrams of (a) iron–carbon, (b) iron–chromium and (c) iron–nickel–chromium are shown in Fig. 3.23. Austenitic stainless steel is basically an iron–chromium–nickel alloy group with varying amounts of carbon and other elements to provide special properties. Adding nickel to iron–chromium alloys widens the range in which austenite exists. No alpha ferrite is found in these steels, but a small amount of delta ferrite, which forms at elevated temperatures, may remain in a metastable form after cooling to room temperature.

Ferrite, which forms in a weld, exhibits a relatively high solubility for sulphur, phosphorus and other constituents known to cause hot cracking. Consequently, welds that contain too little delta ferrite are susceptible to hot cracking. However, ferrite, a BCC structure, experiences a ductile-to-

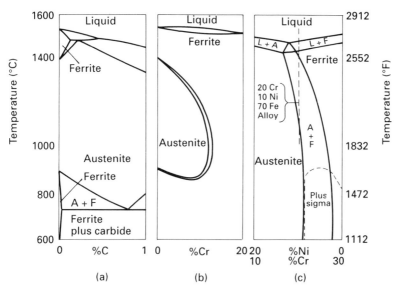

3.23 Phase diagrams: (a) iron–carbon; (b) iron–chromium; (c) iron–nickel–chromium (source: DeLong, W.T., 1974, *Welding Journal*, **54** (7)).

brittle transition, which causes loss of ductility and subsequent cracking in the stainless steel welds at low (e.g. cryogenic) temperatures. Moreover, ferrite has a high solubility for chromium and therefore offers a potential site for the nucleation of sigma phase. Sigma phase is a brittle, high chromium content, intermetallic compound that forms after long exposure to elevated temperatures, and that causes a significant loss of ductility in stainless steel welds subjected to those conditions. Consequently, too much ferrite can produce a cracking problem in service under certain conditions, whereas too little can result in solidification cracking (hot cracking), and in cracking of the weld just as it solidifies.[29, 30] Ferrite number (FN), nearly equivalent to the percentage of ferrite, is a common measure of the ferrite content in the cooled weld metal microstructure.

The ferrites in austenitic stainless steel welds have four types of weld microstructure: vermicular, acicular, lathy and lacy.[31, 32] These four structures are shown in Fig. 3.24. Often the weld metal has no single representative microstructure, but rather a combination of structures. These structures are influenced by local variations in cooling rates and chemical composition.[31] According to David, the vermicular morphology is most commonly found in austenitic stainless steel welds in the range of ferrite numbers from 5 to 15, and is characterized by an aligned skeletal network. The acicular morphology is characterized by the random arrangement

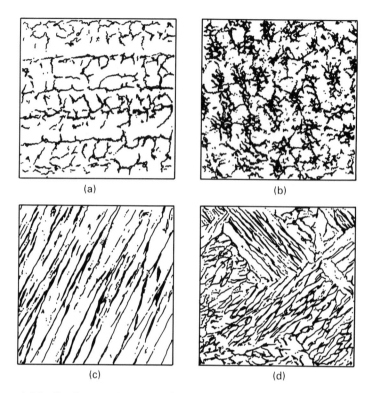

3.24 Ferrite morphologies: (a) vermicular; (b) lacy; (c) lathy; (d) acicular ×750 (source: TVA report, 'Weldability of nuclear grade stainless steels', contract no. TVA 64853A, September 1986).

of needle-like ferrite in an austenite matrix. The lathy morphology is characterized by the ferrite laths running across to the opposite grain boundary.[32] The ferrite has no relationship to the dendrites, but may be related to the grain and its growth direction. The lacy morphology is characterized by long columns of interlaced ferrite network oriented along the growth direction in an austenite matrix.

A model for the different solidification modes and resulting structures is shown in Fig. 3.8.[33] According to Suutala, in welds where the delta ferrite has vermicular structure and is located at the cell boundaries, the austenite is the primary solidifying phase (as shown in Fig. 3.25(a) and (b)). For solidification of welds in which vermicular and lathy structure appear, the inverse occurs, such that the delta ferrite is the first to solidify and transforms to austenite either by an acicular (Fig. 3.25(c)) or equiaxial mechanism (Fig. 3.25(d)), depending on the supercooling of the delta ferrite. In welds in which the lathy structure is dominant, only ferrite

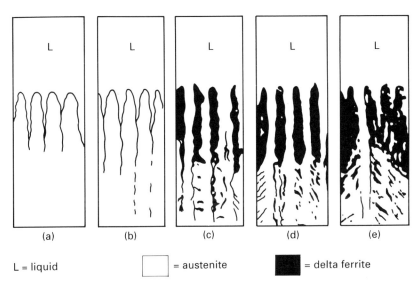

3.25   Proposed solidification model for austenitic and austenitic-ferritic weld metals: (a) fully austenitic; (b) austenite solidifies first; (c) delta ferrite solidifies first then a phase transformation; (d) equi-axial growth of delta ferrite; (e) weld metal solidifies to delta ferrite and austenite form through a solid state transformation (source: Suutala, N., Takalo, T., and Moisio, T., 1979, *Met. Trans. A* **10A** 512).

solidifies directly from the melt, and austenite forms through a solid-state transformation (Fig. 3.25(d)).

Austenitic stainless steel welds commonly have a structure composed of vermicular delta ferrite, which is the primary solidifying phase.[34]

The composition (the amount of alloying element added) is important. The tendency to form delta ferrite at the solution treatment temperature will decrease if the element added is an austenite former, or increase if it is a ferrite former.[30] Table 3.4 lists the elements that influence austenite and ferrite formation.[29] Using empirical constitution diagrams proposed by Schaeffler and DeLong,[29,30] the as-welded ferrite content can be predicted from the chemical composition of the deposited weld metal. The balance between ferrite-forming elements (expressed as chromium equivalent, Cr_{eq}), and austenite-formers (expressed as a nickel equivalent, Ni_{eq}), controls the structure of the weld metal. The influence of the residual elements is factored into the equations for Cr_{eq} and Ni_{eq}. Chemical compositions of welding electrodes generally aim at a balance giving 3–10% ferrite.

Table 3.4. Influence of residual elements on the phase balance in stainless steels

Austenite formers	Ferrite formers
Manganese	Chromium
Cobalt	Molybdenum
Nickel	Aluminium
Nitrogen	Phosphorus
Carbon	Vanadium
	Titanium
	Columbium
	Tungsten
	Zirconium
	Silicon

3.6.2   Sensitization

Sensitization, the susceptibility of stainless steels to intergranular corrosion (localized attack occurring along the grain boundaries), can occur after thermal processing. In many cases, sensitization results from welding.[34] Sensitization in austenitic stainless steels can occur in three ways:

- Ordinary stainless that has been as-welded.
- Ordinary stainless that has been improperly heat-treated.
- Stabilized stainless that has been improperly heat-treated.

Limiting carbon content and temperature minimizes sensitization during welding. Carbon content is the most important factor in determining the susceptibility of austenitic stainless steels to sensitization. The carbon content can be limited by changing to L-grade (carbon below 0.03%) material. Doing so usually avoids sensitization.[34]

The addition of carbon atoms to the alloy of iron and chromium is limited by the carbon's solubility (the quantity of solute that dissolves in a given quantity of solvent to form a saturated solution). The structure can accept only a certain amount of carbon in solution (similar to the way only a limited amount of salt can be dissolved in water).[2] The excess carbon atoms combine with the chromium atoms and form chromium carbides (or other metallic carbides, depending on the element involved), when sufficient heat is applied for the carbide precipitation to occur.[2]

A solid solution is a homogeneous crystalline phase composed of several distinct chemical species. Each atom type normally occupies the lattice points at random and exists in a range of concentrations.[2] Under an optical microscope, a solid solution cannot be distinguished from a pure element. Austenitic stainless steel is a solid solution of chromium and

Cr depleted zone   Grain boundary

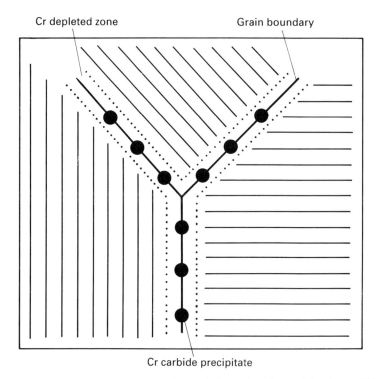

Cr carbide precipitate

3.26 Schematic illustration of a grain boundary in sensitized type 304 stainless steel (data from Sindo Kou, *Welding Metallurgy*, copyright © 1987 John Wiley & Sons. Reprinted by permission of John Wiley & Sons, Inc.).

nickel in iron. If, however, the solubility for a particular element is exceeded (within the matrix), a second phase forms, as happens in the formation of chromium carbides in austenitic stainless steel, which is therefore dictated by the phase relations.[2]

The chromium carbides tend to nucleate and grow at the grain boundaries, as shown in Fig. 3.26. Depending on the amount of chromium depletion, precipitation of the chromium carbides at the grain boundaries can cause the steel to be susceptible to intergranular corrosion.

Bain *et al.* developed a model for intergranular corrosion. This model attributes sensitization to two events happening simultaneously.[35] The chromium carbides precipitate at grain boundaries and as the carbides form, they deplete chromium in the adjacent region. A minimum chromium level (approximately 12% Cr) is required to maintain a protective passive film on stainless steels.

Bain's theory contends that in the temperature range of 500–900 °C (932–1652 °F):

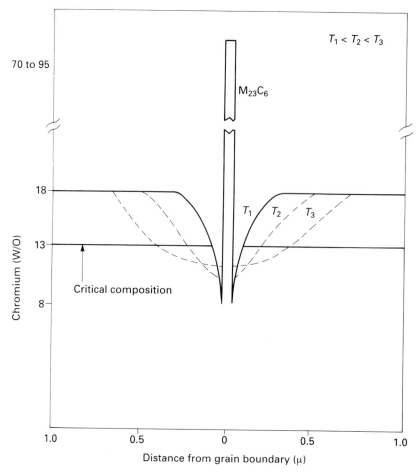

3.27 Schematic diagram of the chromium profile in the vicinity of a grain boundary as a function of time (source: Povich, M.S., 1978, 'Low temperature sensitization of type 304 stainless steel', *Corrosion* **34**, p. 269, reproduced with permission).

- Complex chromium carbide formation (complex in terms of nucleation and growth mechanism) is favourable.
- This carbide precipitation takes place (nucleates and grows) most rapidly at the grain boundaries.

This chromium profile is shown in Fig. 3.27. The chromium-rich carbide particle may contain up to 90–95% chromium, while the bulk alloy may be only 18–20% chromium. The region immediately adjacent to the carbide particle is depleted of chromium. This depletion reduces the local

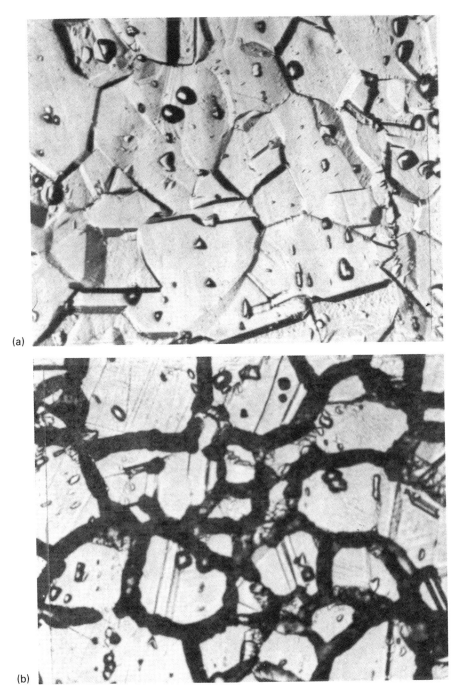

(a)

(b)

3.28 Austenitic stainless steel subjected to oxalic acid etch test (ASTM A262): (a) grain boundaries are not attacked; (b) grain boundaries are preferentially attacked (reproduced with permission from ASTM).

chromium concentration to below the level (approximately 12%) required to provide protection by passivation.

The sensitizing temperature range is between 430–870 °C (800 and 1600 °F). When the austenitic stainless steels are heated or cooled (slowly) through this range, carbon is precipitated from solid solution, and the carbon combines with chromium to form chromium carbides. Usually, the welding procedures limit the maximum interpass temperature (temperature between passes) to 180 °C (350 °F). This temperature serves as a practical limit in minimizing sensitization. The chromium carbide formed during heating reduces corrosion resistance.

A method for testing for sensitization is described in ASTM A262 *Standard Practices for Detecting Susceptibility to Intergranular Attack in Austenitic Stainless Steels*. Figure 3.28 shows the resulting structure using the screening test (practice A-oxalic acid test) in A262. This test, using an oxalic acid etch, reveals the presence or absence of carbides along the grain boundaries. Figure 3.28(b) shows the resulting structure, with the presence of carbides along the grain boundary indicating a sensitized structure.

3.7   A review of nonferrous metals and alloys

Nonferrous metals and alloys include all the metals that are not iron-based. They consist of many commonly used metals, such as copper, nickel and aluminium, as well as exotic metals such as titanium.

3.7.1   Copper and copper alloys

Copper as a metal and alloy was first produced about 4000 BC. The Bronze Age preceded the Iron Age. Copper conducts heat and electricity very well, and has good corrosion resistance.

There are basically two types of (pure) copper. These are electrolytic tough pitch copper (ETP, C11000), and phosphorus deoxidized coppers (DHP, C12200 or C12000). Copper alloys are composed primarily of the brass and bronzes. A brass is a copper with zinc as the primary alloying element. A bronze is a copper with other elements, usually tin, silicon, aluminium or nickel.

Most brasses contain between 5% and 45% zinc. The colour of the brass changes as copper decreases and zinc increases, and a little lead in the brass sometimes appears, which improves machinability.

Copper and copper alloys (Table 3.5) are susceptible to corrosion by ammonia and amines. Brasses are susceptible to stress corrosion cracking. It must be emphasized that corrosion that is not related to MIC does exist. Copper alloys were once thought to be 'immune' to MIC, but we now know that is not true.

Table 3.5. The compositions of some copper alloys

	UNS	wt%					
		Cu	Zn	Sn	Al	Ni	Other
Copper	C11000	99.9	–	–	–	–	–
Red brass	C23000	85	15	–	–	–	–
Yellow brass	C27000	69	31	–	–	–	–
Admiralty	C44300	72	27	1	–	–	0.1
Phosphor bronze	C52400	90	–	10		–	0.3 P
Aluminium bronze	C61400	90	0.2	–	7	–	3 Fe
90%Cu–10% Ni	C60600	86	1.0	–	–	10	1.5 Fe
70%Cu–30% Ni	C71500	68	1.0	–	–	30	1 Fe

There are four general categories of bronzes:[36]

- Phosphor bronze.
- Silicon bronze.
- Aluminum bronze.
- Cupro-nickels.

3.7.2 Aluminium and aluminium alloys

Aluminium was first produced by refining the principal ore, bauxite, in 1825. Aluminium is considered a reactive alloy.[37] It exhibits an aluminium oxide film that is relatively stable and protects it from corrosion.

Aluminium has good malleability and formability, high corrosion resistance, high electrical and thermal conductivity, and is nonsparking. Aluminium's density is about one-third that of steel's. As a result, many aluminium alloys have better strength-to-weight ratios than steel, and are often chosen for this reason.

Pure aluminium's tensile strength is about 90 MPa (13 000 psi). Through alloying and heat treatment, aluminium can achieve a tensile strength of up to 690 MPa (100 000 psi). Temper designations are listed in Table 3.6.

The good corrosion resistance of aluminium is due to the formation of a self-protecting, thin, invisible oxide film that forms immediately on exposure to the atmosphere.

Anodizing is a common process for aluminium. Anodizing forms a conversion coating on the metal surface by anodic oxidation. The anodizing process forms a relatively thick oxide coating on the surface by passing an electric current through the metal. This surface must be sealed to prevent staining and absorption.

Aluminium alloys are identified by four digits (Table 3.7) and a temper designation. The major alloying element is identified by the first digit.

Table 3.6. Temper designations to identify conditions

F	As fabricated
O	Annealed and recrystallized
H	Strain-hardened
W	Solution heat-treated
T	Thermally treated

Table 3.7. Aluminium alloy designations

Alloy	Description
1XXX	Commercially pure aluminium, 99.00% aluminium
2XXX	Copper. These are age-hardenable and include some of the highest strength aluminium alloys, such as Alloy 2024
3XXX	Manganese. These are not heat-treatable, and have good formability, good corrosion resistance and weldability
4XXX	Silicon. These are not heat-treatable, and have excellent castability and corrosion resistance
5XXX	Magnesium. These are not heat-treatable and have good weldability, corrosion resistance and moderate strength
6XXX	Silicon–magnesium. These are artificially aged. They have excellent corrosion resistance and workability. Alloy 6061 is commonly used for structural applications
7XXX	Zinc. These develop the highest tensile strengths, but are susceptible to stress corrosion cracking. They are heat-treatable
8XXX	Other elements

3.7.3   Nickel and nickel alloys

Nickel was discovered in 1751. It is a major alloying element in stainless steels and nickel-based alloys, which are also called superalloys. Major nickel alloys are listed in Table 3.8.

Nickel is noted for good corrosion resistance and is particularly effective in oxidizing environments. It forms tough, ductile solid–solution alloys with many metals. Its mechanical properties are similar to those of mild steel. In addition, it retains its strength at elevated temperatures while also maintaining ductility and toughness at low temperatures.

Nickel–copper alloys such as Monel^{TM} have excellent corrosion resistance to a wide variety of environments and may be used at temperatures of up to 800 °C (1500 °F).

Nickel–silicon alloys such as Hastelloy^{TM} D are strong, tough and extremely hard. They have excellent corrosion resistance in sulphuric acid at high temperatures.

Table 3.8. Nickel and its alloys

| Alloy | UNS | wt% | | | | | |
		Ni	Cr	Fe	Mo	C	Other
200	N02200	99	–	0.4	–	0.15	–
400	N04400	63–70	–	1.0–2.5	–	0.3	23–34 Cu
600	N06600	72	14–17	6–10.0	–	0.15	–
800	N08800	32	21	bal.			
C-276	N10276	bal.	14.5–16.5	4–7	15–17	0.02	3.0–4.0 W

Nickel-chromium alloys, such as Inconel™ 600, and nickel–chromium–iron alloys, such as Inconel™ 800, combine the corrosion resistance, strength and toughness of nickel with the high temperature oxidation resistance of chromium. (Monel, Hastelloy and Inconel are trademarks of INCO.)

Nickel–molybdenum–iron alloys such as Hastelloy™ B have high corrosion resistance to hydrochloric, phosphoric and other nonoxidizing acids.

Nickel–chromium–molybdenum–iron alloys such as Hastelloy™ C or C-276 have high corrosion resistance to oxidizing acids, good high temperature properties, and are resistant to oxidizing and reducing atmospheres up to 1100 °C (2000 °F).

3.7.4   Titanium alloys

Titanium was discovered in 1791, but was not commercially available until the 1950s. Its most important features are its high strength-to-weight ratio and its excellent corrosion resistance. Its corrosion resistance is due to a continuous, tough, adherent, passive oxide film TiO_2 that is stable in a wide pH range.[38] The corrosion resistance is considered outstanding to:[39]

- Seawater and other chloride salt solutions.
- Hypochlorites and wet chlorine.
- Nitric acid, including fuming acid.

The density of titanium is about 4430 kg/m³ (0.16 lb/in³), compared to steel's 7750 kg/m³ (0.28 lb/in³), producing the excellent strength-to-weight ratio. Titanium alloys have been developed for service with corrosion resistance up to 540 °C (1000 °F) with long exposure times. Some titanium alloys are listed in Table 3.9.

Table 3.9. Titanium alloys

Alloy	Designation	UNS	wt%
Pure	Grade 1	R50250	0.20 Fe, 0.18 O, bal. Ti
Pure	Grade 2	R50400	0.30 Fe, 0.25 O, bal. Ti
Ti–6Al–4V	Grade 5	R56400	5.5–6.5 Al, 0.40 Fe, 0.20 O, 3.5–4.5 V, bal. Ti

3.8   Heat treatment of nonferrous alloys and stainless steels

Nonferrous alloys are heat-treated differently from ferrous and some stainless steel alloys because they do not undergo phase transformations as steels do. The hardening and strengthening are produced by different mechanisms.

Aluminium alloys capable of hardening by heat treatment, as well as copper alloys, martensitic and precipitation-hardening stainless steels, are hardened by precipitation hardening. Precipitation hardening, also called age hardening, means hardening by the precipitation of a different phase (as small particles called precipitates). This phase is derived from a supersaturated solid solution. The particles are dispersed in the metal matrix.

Some alloys can be heat-treated and their properties modified by two different methods: solution treatment and precipitation hardening.

Precipitation hardening of steel to form tempered martensite is a different process, although similar. Precipitation hardening may occur by two different heat treatments.

- Solution heat treatment. In this process, a metallurgist forms a supersaturated or metastable solid solution. The alloy is heated to the appropriate temperature, held to dissolve the precipitates, and quenched.
- Precipitation heat treatment. For some alloys, a second treatment is needed. The alloy is reheated to an intermediate temperature (below the solution heat treatment temperature), held at this temperature to allow precipitation to occur, and allowed to cool. The rate of cooling is often not an important consideration.

3.9   Metallurgical factors affecting MIC

In addition to mechanical influences, microbiological influences also affect corrosion, as discussed in Chapter 2. In many cases, these influences are associated with colonies of organisms, scale, debris and exopolymers forming local deposits on a metal surface. The deposits often initiate pits underneath on the surface of the metal. The pits grow and propagate, leading to corrosion of the metal.

Table 3.10. A summary of materials and
fabrication affecting MIC

Metallurgy
  Alloy composition
  Trace elements
  Inclusions
  Sulphur content
  Passive films
  Oxide films
  Heat treatment
    Annealing
    Stress relief
Welding
  Heat input
  Joint design and fitup
  HAZ
  Filler metal
  Oxide films
Residual stress
Galvanic effects
  Magnetic properties
Surface effects
  Roughness
  Machining
  Oxide films
  Pickling
  Gouges
  Contamination
  Electron donors and electron acceptors
  Crevices
  Flow characteristics

Researchers have studied the adhesion of microorganisms to metal surfaces and the development of surface colonies.[40, 41] When biofilms form on metal surfaces, a number of factors on that surface affect this process of adherence. These factors, with examples, include:

- Type of metal, e.g. copper-based versus iron-based.
- Alloy content of metal, e.g. sulphide inclusions.
- Surface condition, e.g. rough versus smooth.
- Fabrication conditions, e.g. oils, residues.
- Microstructure, e.g. heat treatment, welding.
- Combinations of metals, e.g. weldments, dissimilar metals.

Unless we understand the complexities of the metallurgical influences and their implications, we will continue to have unidentified variables, which inevitably result in poor laboratory experiments (i.e. laboratory work that refers to 'a metal coupon' and wasted research dollars). Table 3.10 summarizes some of the metallurgical factors that affect MIC.

3.9.1   The effect of stress

Under some conditions, a metal will show a higher corrosion rate when under tensile stress.[42] Of more concern is the cracking of a metal that may occur when a susceptible metal is under the combined effects of tensile stress and corrosion. This may produce a brittle failure of a normally ductile metal. Residual stress is usually the result of cold working and may affect corrosion resistance.

Residual stresses are a concern in relation to susceptibility to MIC, because stresses are often introduced during fabrication or in service. Stein theorizes that MIC in austenitic stainless steel is associated with residual stresses.[43]

3.9.2   The effect of alloying elements

The properties of an alloy are made of a combination of the physical and chemical characteristics. That combination depends in part on the microstructure, comprising characteristics derived from the amount, distribution, shape and so forth of the phases in the alloy. Multiple phases within an alloy may behave differently, because of differences in electro-chemical properties.[44]

Szklarska-Smialowska discusses the effects of alloying elements on pitting.[45] Many researchers have investigated this subject. Szklarska-Smialowska notes that, according to Kolotyrkin, the pitting susceptibility of steels decreases with increasing levels of nitrogen, nickel and especially chromium and molybdenum.[46] Szklarska-Smialowska observes that for stainless steels, the addition of molybdenum to austenitic and ferritic stainless steels improves their corrosion resistance and passive properties.[46] This concept of pitting and pitting potential, the critical potential above which pits nucleate and develop, is discussed in Section 4.5.

The effects of alloying elements are a concern in relation to susceptibility to MIC, since the surface inhomogeneity may contribute to the initial adhesion and subsequent corrosion. For example, Kearns and Borenstein state that austenitic stainless steel welds with filler metal compositions matching the base metal have a lower corrosion resistance than fully annealed base metal. This is due to the lack of homogeneity and the microsegregation of chromium and molybdenum.[47] Localized attack is more likely at chemically depleted regions.

3.9.3   The effect of oxide film thickness

The protective nature of the passive oxide film is discussed in Section 4.5. The influence of passive films and oxide film thicknesses is important to understanding pitting mechanisms.[48] The surface conditions (e.g. heat tint, scratches and gouges) of a metal influence the susceptibility to corrosion, and to MIC as well. This is discussed in Chapter 4.

3.9.4   The effects of cold work

Studies have been done on the effect of stress, such as cold work on pitting corrosion.[49] Szklarska-Smialowska reports that the studies gave mixed results, with cold working influencing whether the number of pits per unit area increased, decreased, or remained unchanged. As above, cold working of a metal influences its susceptibility to corrosion and MIC.

3.9.5   The effects of heat treatment

Most of the properties of a fabricated component are related to its heat treatment.[50] As described earlier, annealing of metal alloys is usually done to produce either a homogenized structure or to remove residual stress and cold work.

Pitting due to MIC at or adjacent to weldments of austenitic stainless steels is common. Both the austenite and ferrite phases may be susceptible. My own research has shown that for austenitic stainless steels solution annealing and pickling may produce welds that are less susceptible.[51]

The effects of sensitization heat treatments are a concern in relation to susceptibility to MIC, because it lowers corrosion resistance. Videla *et al.* found that sensitizing heat treatments of 304 and 410 stainless steel reduced the pitting corrosion resistance in the presence of SRBs and aggressive ions.[52]

Suggested reading

1 ASM Metals Handbook, 9th ed.:
   - Vol. 1: *Properties and Selection: Irons and Steels*, 1978.
   - Vol. 2: *Properties and Selection: Nonferrous Alloys and Pure Metals*, 1979.
   - Vol. 3: *Properties and Selection: Stainless Steels, Tool Materials and Special Purpose Metals*, 1980.
   - Vol. 4: *Heat Treating*, 1985.
2 United States Steel Corp., 1950, *The Making, Shaping and Treating of Steel.*
3 Peckner, D., and Bernstein, I.M., eds., 1977, *Handbook of Stainless Steels*, McGraw-Hill, New York.
4 Woldman, N.E., 1979, *Engineering Alloys*, ASM International, Metals Park, OH.
5 Kalpukjian, S., 1989, *Manufacturing Engineering and Technology*, Addison-Wesley, Reading, MA.
6 Miller, J.D.A., 1970, *Microbial Aspects of Metallurgy*, American Elsevier, New York.

7   Boyer, H.E., 1984, *Practical Heat Treating*, ASM International, Metals Park, OH.
8   Brooks, C.R., 1984, *Heat Treatment*, ASM International, Metals Park, OH.
9   American Welding Society, 1976, *Welding Handbook*, 7th ed., 5 vols, Miami, FL.
10 American Welding Society, 1987, *Welding Handbook*, 8th ed., 3 vols., Miami, FL.
11 Kou, S., 1987, *Welding Metallurgy*, Wiley Interscience, New York.
12 Lancaster, J.F., 1980, *Metallurgy of Welding*, 3rd ed., George Allen & Unwin, London.
13 Dowling, N.J., Mittleman, M., and Danko, M., ed. 1990, *Microbially Influenced Corrosion and Biodeterioration*, National Association of Corrosion Engineers, Houston, TX.

References

1   Van Vlack, L.C., 1975, *Elements of Materials Science and Engineering*, 3rd ed. Addison-Wesley, Reading, PA, p. 303.
2   Parr, J.G., and Hanson, A., 1965, *An Introduction to Stainless Steel*, ASM International, Metals Park, OH, p. 19.
3   Fontana, M., 1986, *Corrosion Engineering*, McGraw-Hill, New York, p. 393.
4   ASM Metals Handbook, 1980, 9th ed., Vol. 3, *Stainless Steels, Tool Materials and Special Purpose Metals*, ASM International, Metals Park, OH, p. 7.
5   Peckner, D., and Bernstein, I.M., 1977, *The Handbook of Stainless Steels*, McGraw-Hill, New York, p. 1-4.
6   Efird, K.D., and Moller, G.E., 1979, *Materials Performance*, **18**(7) 34.
7   ASM, Metals Handbook, 1978, 9th ed., Volume 1, *Properties and Selection: Irons and Steels*, ASM International, Metals Park, OH.
8   Avner, S.H., 1974, *Introduction to Physical Metallurgy*, McGraw-Hill, New York, p. 265.
9   Avner, S.H., 1974, *Introduction to Physical Metallurgy*, McGraw-Hill, New York, p. 270.
10 Avner, S.H., 1974, *Introduction to Physical Metallurgy*, McGraw-Hill, New York, p. 273.
11 Kalpukjian, S., 1989, *Manufacturing Engineering and Technology*, Addison-Wesley, Reading, MA.
12 AWS, 1968, *Introduction to Welding Metallurgy*, American Welding Society, Miami.
13 Allen, D., 1969, *Metallurgy Theory and Practice*, American Technical Publishers, Homewood, IL, p. 591.
14 Allen, D., 1969, *Metallurgy Theory and Practice*, American Technical Publishers, Homewood, IL, p. 592.
15 AWS, 1968, *Introduction to Welding Metallurgy*, American Welding Society, Miami, p. 3.
16 AWS, 1968, *Introduction to Welding Metallurgy*, American Welding Society, Miami, p. 40.
17 AWS, 1968, *Introduction to Welding Metallurgy*, American Welding Society, Miami, p. 12.

18 AWS, 1968, *Introduction to Welding Metallurgy*, American Welding Society, Miami, p. 44.

19 Kalpukjian, S., 1989, *Manufacturing Engineering and Technology*, Addison-Wesley, Reading, MA, p. 872.

20 Kalpukjian, S., 1989, *Manufacturing Engineering and Technology*, Addison-Wesley, Reading, MA, p. 870.

21 AWS, 1968, *Introduction to Welding Metallurgy*, American Welding Society, Miami, p. 7.

22 Kalpukjian, S., 1989, *Manufacturing Engineering and Technology*, Addison-Wesley, Reading, MA, p. 820.

23 Kalpukjian, S., 1989, *Manufacturing Engineering and Technology*, Addison-Wesley, Reading, MA, p. 823.

24 Kalpukjian, S., 1989, *Manufacturing Engineering and Technology*, Addison-Wesley, Reading, MA, p. 824.

25 Kalpukjian, S., 1989, *Manufacturing Engineering and Technology*, Addison-Wesley, Reading, MA, p. 827.

26 AWS, 1968, *Introduction to Welding Metallurgy*, American Welding Society, Miami, p. 110.

27 AWS, 1968, *Introduction to Welding Metallurgy*, American Welding Society, Miami, p. 16.

28 AWS, 1968, *Introduction to Welding Metallurgy*, American Welding Society, Miami, p. 118.

29 Lippold, J.C., and Savage, W.F., 1979, *Welding Journal*, **59**(12) 362s.

30 Peckner, D., and Bernstein, I.M., 1977, *The Handbook of Stainless Steels*, McGraw-Hill, New York, p. 14-4.

31 David, S.A., 1981, 'Ferrite morphology and variations in ferrite content in austenitic stainless steel welds', *Welding Journal*, 63s.

32 Lundin, C.D., Osorio, V., Lee, C.H., Menon, R., 1986, 'The weldability of nuclear grade stainless steels', TVA Contract No. TVA64853A.

33 Suutala, N., Takalo, T., Moisio, T., 1979, *Met. Trans. A.* **10A** 512.

34 ASM Metals Handbook, 1983, 9th ed., Vol. 6, *Welding Brazing, and Soldering*, ASM International, Metals Park, OH, p. 117.

35 Bain, E.C., Aborn, R.H., and Rutherford, J.B., 1933, 'The nature and prevention of intergranular corrosion in austenitic stainless steels', *Trans. A.S.S.T.*, **21** 481.

36 Dillon, C.P., 1986, *Corrosion Control in the Chemical Process Industries*, McGraw-Hill, New York, p. 111.

37 NACE, 1984, *Corrosion Basics, An Introduction*, National Association of Corrosion Engineers, Houston, TX, p. 75.

38 ASM Metals Handbook, 1987, Vol. 13, *Corrosion*, ASM International, Metals Park, OH, p. 669.

39 NACE, 1984, *Corrosion Basics, An Introduction*, National Association of Corrosion Engineers, Houston, TX, p. 78.

40 Ford, T., and Mitchell, R., 1991, 'The ecology of microbial corrosion', in *Advances in Microbial Ecology*, Marshall, K.C., ed., Plenum Press, New York, p. 233.

41 Costerton, J.W., Geesey, G.G., and Cheng, K.J., 1978, 'How bacteria stick', *Scientific American*, No. 238, 86.

42 NACE, 1984, *Corrosion Basics, An Introduction*, National Association of Corrosion Engineers, Houston, TX, p. 12.
43 Stein, A.A., 1991, Proceedings *Corrosion/91*, paper no. 107, National Association of Corrosion Engineers, Houston, TX.
44 NACE, 1984, *Corrosion Basics, An Introduction*, National Association of Corrosion Engineers, Houston, TX, p. 53.
45 Szklarska-Smialowska, Z., 1986, *Pitting Corrosion of Metals*, NACE, Houston, TX, p. 143.
46 Szklarska-Smialowska, Z., 1986, *Pitting Corrosion of Metals*, NACE, Houston, TX, p. 147.
47 Kearns, J. and Borenstein, S.W., 1991, 'Microbially influenced corrosion testing of welded stainless steel alloys for nuclear power plant service water systems', *Corrosion/91*, paper no. 279, National Association of Corrosion Engineers, Houston, TX.
48 Szklarska-Smialowska, Z., 1986, *Pitting Corrosion of Metals*, National Association of Corrosion Engineers, Houston, TX, p. 163.
49 Szklarska-Smialowska, Z., 1986, *Pitting Corrosion of Metals*, National Association of Corrosion Engineers, Houston, TX, p. 176.
50 NACE, 1984, *Corrosion Basics, An Introduction*, National Association of Corrosion Engineers, Houston, TX, p. 57.
51 Borenstein, S.W., 1991, 'Microbiologically influenced corrosion of austenitic stainless steel weldments', *Materials Performance*, **30**(11) 52.
52 Videla, H.A., deMele, M.F.L., Moreno, D.A., Ibars, J., and Ranninger, C., 1991, 'Influence of Microstructure on the Corrosion Behaviour of Different Stainless Steel', *Corrosion/91*, paper no. 104, National Association of Corrosion Engineers, Houston, TX.

Chapter 4

Electrochemistry

Electrochemistry is the science of chemical changes that accompany the passage of electric current. Piron defined electrochemistry as a chemical science and technology, concerned with the properties of ions in solution and their reactions at metal–solution interfaces.[1] Electric conduction occurs through the motion of charged particles, of either electrons or ions. Current passes through an electrolyte and causes chemical reactions at the electrodes. A galvanic cell, or battery, is an example of a chemical reaction arranged to produce electrons. The reverse, using electricity to produce a chemical reaction, is termed electroplating, and is also an example of electrochemistry. The electrochemical nature of corrosion offers a method of determining the corrosion rate. This chapter covers electrochemistry's relationship with the corrosion of metals. It describes how microorganisms can influence electrochemical reactions and corrosion and how electrochemical methods may be used to investigate such processes.

Electrochemistry can be discussed in terms of thermodynamics and kinetics. Thermodynamics will tell us what *can* happen, kinetics will tell us what *will* happen.[2] Thermodynamics relates to the conversion of heat into energy, while kinetics relates to the study of reaction rates and the factors that affect those rates.

4.1   Glossary

The following are frequently used terms in electrochemistry and corrosion.

4.1.1   Corrosion-related terms

- **Corrosion**: the deterioration of a substance (usually a metal) or its properties, due to a reaction with its environment.
- A **chemical reaction**: a reaction that does not involve the transfer of electrons.
- An **electrochemical reaction**: a chemical reaction involving the transfer of electrons.

- **Electrode**: a metal in contact with an electrolyte at the location where the current either enters or leaves the metal to enter the solution.
- **Electrolyte**: a medium that conducts electric current by ionic movement.
- **Anions**: negatively charged ions attracted to the anode.
- **Cations**: positively charged ions attracted to the cathode.
- **Anode**: the portion of the metal where current is discharged and corrosion occurs.
- **Cathode**: the metal surface where electrons from the anode (by way of the electronic path) react with the electrolyte.
- The **pH**: measurement of the hydrogen ion concentration. It generally denotes the degree of acidity or basicity of a solution:

$$pH = -\log(H^+)$$

- **Oxidation**: this process occurs when electrons are lost from an atom or compound. Oxidation decreases the negative charge (or increases the positive charge). Oxidation occurs at the anode.
- **Reduction**: this process occurs when electrons are added to an atom or compound. Reduction increases the negative charge (or decreases the positive charge). Reduction occurs at the cathode.
- **Half cell reactions**: two separate reactions, anodic and cathodic, occurring simultaneously, owing to the immersion of an electrode in an electrolyte designed to measure a single electrode's potential.
- **Polarization**: the flow of potential because of current flow.
- **Concentration polarization**: the retarding of an electrochemical reaction as a result of concentration changes in the solution adjacent to the metal surface.
- **Cathodic polarization**: when a cathodic reaction product such as, for example, hydrogen, is not removed.
- **Activation polarization**: the retarding actions that are inherent to the reaction itself.
- **Passivity**: the loss of chemical reactivity. The exact nature or cause of these effects is not completely understood.
- **Cathodic protection**: the reduction or elimination of corrosion by making the metal a cathode, by means of an impressed direct current or attachment to a sacrificial anode.
- **Noble**: a chemically unreactive metal (such as gold).

4.1.2 Electrochemistry-related terms

- **Potential**: the relative voltage at a point with respect to a reference point in the same circuit.
- **Open cell potential** (OCP), also called E_{corr}: the potential of an

electrode measured with respect to a reference electrode or another electrode when no current flows to or from it.

- **Current** (I): the flow rate of charged electrons. An ampere is the flow rate of 1 coulomb (6.2×10^{18} electrons) per second.
- I_{corr}: the current density at the free corrosion potential.
- The **Tafel equation**: based on work by Tafel in 1905, who found that the current I equivalent to the rate of a single electrode reaction is related to the potential of the metal by:[3]

$$E = a + b \log I$$

- **Resistance** (R or r): the opposition to current flow through a material. It is a material-dependent property. The ohm is the unit of resistance.
- **Ohm**: the resistance of a conductor across which there is a potential drop of 1 volt when 1 ampere of current flows through it.
- **Electrical resistivity** (ρ): the electrical resistance offered by a material to the flow of current, multiplied by the cross-sectional area of current flow per unit length of current path.
- **Conductivity**: the reciprocal of resistivity.
- **Ohm's law**: the fundamental relationship of current, voltage and resistance in an electrical circuit ($V = IR$).
- **Electromotive force** (EMF): the force that causes current to flow in a circuit, equivalent to potential difference.
- **Cyclic anodic polarization curve**: a curve produced by the cyclic potentiostatic polarization test described by ASTM Standard Practice G-1.
- **Critical current density** (i_c): also called the critical anodic current density. This is the maximum anodic current density observed in the active region of a metal that exhibits active passive behaviour.[4]
- **Passive current density** (i_{pass}): current density in the passive range.
- **Primary passivation potential** (E_{pp}): the potential below which no pitting occurs and above which pits already nucleated may grow.[4]
- **Pitting potential** (E_p): denotes the potential above which pits may nucleate and grow but below which it is unlikely for pitting to occur.[4]
- **Repassivation potential** (E_{rp}): the potential where the reverse scan current density equals the forward scan current density, and generally represents the potential at which active pits repassivate.[4]
- **Breakdown potential** (E_b): the potential above which pits nucleate and develop.[4]

4.2   Electrochemical principles

What is the difference between electrochemical and chemical corrosion? Corrosion is a chemical reaction, as in, for example, zinc placed in hydrochloric acid, which produces a rapid reaction, shown as,

$$Zn + 2HCl \rightarrow ZnCl_2 + H_2 \qquad [4.1]$$

An electrochemical reaction is a chemical reaction that involves the transfer of electrons comprising oxidation and reduction reactions. Electrochemical corrosion therefore involves a transfer of electrons, while chemical corrosion does not. Most aqueous corrosion involves the transfer of electrons.

Because of the electrochemical nature of corrosion, electrochemical techniques are useful in evaluating corrosion processes. Localized corrosion, such as pitting corrosion, is an example of a corrosion mechanism often studied that uses electrochemical techniques.

In pitting, the attack focuses on a small area of the metal surface; the rest of the surface may remain unaffected. The pits initiate as small points on the surface and grow with time.

Pitting corrosion, one of the most destructive and insidious forms of corrosion, is difficult to assess and predict satisfactorily.[5] A number of theories exist to explain the initiation of pits on a surface. In all the theories, metal atoms transfer from the metal to the solution (as ions) as part of the corrosion process.

As shown in Fig. 4.1, the cathode consumes electrons where reduction occurs by one reaction, such as:

$$2H^+ + 2e^- \rightarrow H_2 \qquad [4.2]$$

The hydrogen reduction is only one of many possible reduction reactions. Metal oxidizes (corrodes) at the anode by another reaction:

$$M \rightarrow M^{+2} + 2e^- \qquad [4.3]$$

where 'M' is the metal going to a metal ion plus two electrons.

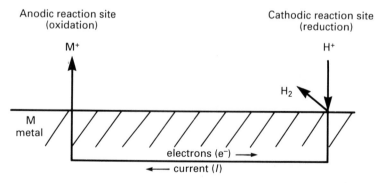

4.1 Corrosion reaction of a single piece of metal (source: Stansbury, E.E., 1983, *Fundamentals of Corrosion*, University of Tennessee).

4.3   Electrochemistry and corrosion

The National Association of Corrosion Engineers (NACE) defines corrosion as 'the deterioration of a material of construction or its properties as the result of exposure to an environment.'[6]

For corrosion to occur in an aqueous environment, four basic elements are required:

- An anode.
- A cathode.
- An electrolyte.
- A current path or circuit.

The corroding metal (or area) is the anode, the noncorroding metal (or area different from the anode) is the cathode, the electrolyte is the liquid in which the metals are in contact and the electrons (and electricity) flow through the path between the two electrodes, the anode, and the cathode.

Understanding rust

The corrosion process can begin when steel is immersed in water containing electrolytes. These are the ions that cause current to flow from the anode to the cathode. Assume that we have a piece of carbon steel immersed in water. The carbon steel surfaces exhibit anodic and cathodic reactions:

$$Fe \rightarrow Fe^{2+} + 2e^-$$ [4.4]

This is an oxidation reaction and occurs at the anode. The iron atoms, Fe, lose electrons and become ferrous ions, Fe^{2+}.

$$2H^+ + 2e^- \rightarrow H_2$$ [4.5]

This is a reduction reaction and occurs at the cathode. The reactions happen simultaneously. The ferrous ions react with water and produce:

$$Fe^{2+} + 2H_2O \rightarrow Fe(OH)_2 + 2H^+$$ [4.6]

The iron ions in water produce ferrous hydroxide. The iron oxides and hydroxides build up on the cathode and form the corrosion product widely known as rust.

Corrosion is a natural process: rust forms on steel that is exposed to air and water. Many chemical compounds disassociate into ions when they are immersed in water and these charged particles form the electrolyte that conducts electric current in an aqueous solution.

Electrochemical reactions

An electrochemical involves oxidation and reduction, as can be illustrated with a dry cell battery, shown in Fig. 4.2. A dry cell battery depends on

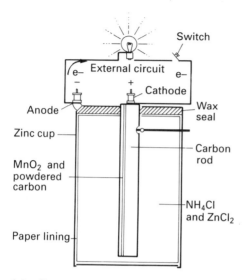

4.2 Cross-sectional view of a typical dry cell (source: NACE, 1984, *Corrosion Basics: An Introduction*, NACE, Houston, TX, reproduced with permission).

corrosion to generate electricity, in a galvanic corrosion process.

In the dry cell, zinc is the anode, and is electrically connected to the graphite cathode in an electrolyte. When a circuit is completed, electrical current flows between the anode and the cathode. The anode corrodes, and produces a cathodic reaction at the graphite cathode. When the zinc is completely corroded (or the electrolyte dries up), the battery is depleted and does not supply any more power.[7]

To review: zinc is the anode, graphite the cathode. HCl and $ZnCl_2$ are the electrolytes in the following reactions:

$$Zn + 2HCl \rightarrow ZnCl_2 + H_2 \qquad [4.7]$$
the full reaction

$$Zn + 2H^+ + 2Cl^- \rightarrow Zn^{2+} + 2Cl^- + H_2 \qquad [4.8]$$
ions reacting

$$Zn + 2H^+ \rightarrow Zn^{2+} + H_2 \qquad [4.9]$$
the net reaction

Zinc is oxidized to zinc ions. Hydrogen ions are reduced to hydrogen gas. Since the chloride ion appears on both sides and does not participate in the reaction, it has been omitted from equation 4.9.

The reaction can be simplified into a cathodic reaction and an anodic reaction. As shown above, the same net reaction occurs (the sum of the half reactions):

$$Zn \rightarrow Zn^{2+} + 2e^- \qquad\qquad\qquad [4.10]$$
oxidation (anodic reaction)

$$2e^- + 2H^+ \rightarrow H_2 \qquad\qquad\qquad\qquad [4.11]$$
reduction (cathodic reaction)

$$Zn + 2H^+ \rightarrow Zn^{2+} + H_2 \qquad\qquad\qquad [4.9]$$
the net reaction

An oxidation reaction means there is an increase in valance, or that electrons are produced. Similarly, a reduction reaction means there is a decrease in valance, or that electrons are used. The oxidation is the anodic reaction, and the reduction is the cathodic reaction. These terms are often used interchangeably. The separate reactions, anodic and cathodic, occur simultaneously and are sometimes referred to as half cell reactions. A half cell reaction is the reaction of a cell consisting of an electrode immersed in an electrolyte designed to measure a single electrode's potential.

Some texts discuss positive and negative flow of current but these terms can be confusing. Conventional current flow is from + to − and will be from the cathode to the anode in a metallic circuit. Electrical current is sometimes called positive flow or positive current. Electron flow is in the opposite direction, and so is described as negative flow or negative current. For clarity's sake, this text uses 'current' to refer to electrical current only; we also refer only to positive flow and positive current.

Polarization

Polarization is the shift of potential as a result of current flow, and is probably the most important concept in corrosion control. As discussed earlier, a half cell reaction is the tendency of a reaction to gain or lose electrons.

Corrosion reactions tend to slow down or speed up (change rate) as a result of what happens at the anode and cathode reactions. Cathodic polarization refers to when a cathodic reaction product (for example, hydrogen), is not removed. If the hydrogen product at the cathode discussed above was not removed by a gas evolving, for example, or by some other reaction involving oxygen, the whole reaction would slow down. This slowing down is the result of cathodic polarization.[8]

This effect can be measured, in terms of the potential of the metal where the reaction is occurring, by measuring the potential of the surface of the more noble metal, the cathode, before the two are coupled (i.e. before the flow of any galvanic current.) When they are coupled, a current flow will eventually occur. The potential will have changed, and will now be closer to the less noble metal. The anode will behave in a similar fashion, and will show a drift in potentials closer to the cathode.

Since there are two types of polarization, concentration and activation, there are two ways in which electrochemical reactions can be slowed down.

Concentration polarization refers to the retarding of an electrochemical reaction as a result of concentration changes in the solution adjacent to the metal surface. This is a diffusion-controlled reaction. Assume that hydrogen is evolving on the surface of a piece of rapidly corroding metal. If the reaction is happening rapidly, we can see that the reaction in the region close to the metal surface will become depleted of hydrogen, since hydrogen is being consumed by the cathodic reaction. The diffusion rate of the hydrogen ions to the metal surface controls the reaction rate.

Activation polarization refers to the retarding actions inherent in the reaction itself. In a similar example, the rate at which hydrogen ions are reduced to hydrogen gas is a function of several factors. The rate depends on the type of metal, the hydrogen ion concentration and the temperature of the system. This rate can vary widely. Activation polarization is usually the controlling factor in corrosion in strong acids, while concentration polarization is usually the controlling factor in dilute acids, aerated water and salt solutions. Knowing the type of polarization is helpful in investigating a corroding system. For example, in a concentration polarization controlled reaction, stirring would increase the corrosion rate.

Ohm's law

Corrosion can be modelled with electrical circuits. The basics of electricity include an understanding of voltage, current, resistance, resistivity and Ohm's law. Voltage is an electromotive force (EMF), or a difference in electrode potentials expressed in volts. We often use the term potentials and voltage interchangeably. The volt is a unit of electromotive force that causes a current of one ampere to flow through the resistance of one ohm. There are several symbols used for voltage: EMF, E and V.

Current is the flow rate of charged electrons. An ampere is the flow rate of 1 coulomb (6.2×10^{18} electrons) per second.

Resistance is the opposition to current flow through a material and is a material-dependent property. The ohm is the unit of resistance and one ohm is the resistance of a conductor across which there is a potential drop of 1 volt when 1 ampere of current flows through it.

The electrical resistivity, ρ, is the electrical resistance offered by a material to the flow of current, multiplied by the cross-sectional area of current flow per unit length of current path (equation 4.12). The resistivities of common materials are shown in Table 4.1.

$$\rho = (RA)/L \qquad\qquad [4.12]$$

Table 4.1. Resistivity of common materials at 20 °C

Material	Resistivity (Ω cm)
Copper	1.72×10^{-6}
Aluminium	3.2×10^{-6}
Iron	$10-12 \times 10^{-6}$
Zinc	5.9×10^{-6}
Soil (varies)	3000–30 000
Seawater (varies)	30
Dry sand	5×10^5

Resistance to current flow is lowest for:

- Low resistivity solutions.
- High conductivity solutions.
- Short path length of current flow.
- Large cross-sectional area of current flow.

Ohm's law is the fundamental relationship of current, voltage and resistance in an electrical circuit.

$$V = IR \qquad [4.13]$$

where V is the voltage in volts, I is the current in amperes and R is the resistance in ohms.

The current flowing in a galvanic cell is proportional to the amount of metal corroding per unit time by Faraday's law:

$$W = ITZ \qquad [4.14]$$

where W is the weight loss, I is the current in amperes, T is the time and Z is the electrochemical equivalent. Faraday's law provides the relationship between the magnitude of the electrical charge (current multiplied by the time) and the quantity of material that reacts at the electrode. This law can provide us with a measure of how much metal is lost during corrosion.

Faraday's Law is also expressed as:[9]

$$i_{corr} t = (nFw)/M \qquad [4.15]$$

where i_{corr} is expressed in amperes, t is time in seconds, nF is the number of coulombs required to convert 1 mole of metal to corrosion product, n being the number of electrons for the metal dissolution reaction and F the Faraday constant (96 480 C/mol), M is the molecular weight in grams, and w is the mass of corroded metal in grams. One ampere flowing for one hour is 3600 coulombs. Industry often uses amperes and hours (A h) instead of coulombs as a measurement. (One faraday equals 26.8 A h.)

For example, the equivalent weight of copper is defined as the weight divided by n, the number of electrons in the reaction. The atomic weight is 63.54 g and for the reaction $Cu^{2+} + 2e^- \rightarrow Cu$, $n = 2$. Thus, the equivalent weight of copper is 63.54 g divided by 2, which equals 31.77 g. For copper reduction, 26.8 A flowing for one hour produces 31.77 g of copper. Note that the same weight can be deposited with a current of 13.4 A for 2 hours, and so on, which is significant, because a very small amperage over many hours or years can cause a sizeable metal loss.[10] Corrosion currents are typically measured in the microamp range.

4.3.1   Forms of corrosion

Classifying corrosion can be difficult. The NACE gives the forms of corrosion as:

- General or uniform corrosion.
- Localized corrosion (or pitting).
- Galvanic corrosion.
- Cracking phenomenon.
- Velocity effects (erosion–corrosion, cavitation, fretting).
- Intergranular corrosion.
- Dealloying.
- High temperature corrosion.

The individual phenomena of each form of corrosion are shown in Fig. 4.3.

General corrosion

General corrosion is usually the regular loss of a small amount of metal over a large surface. Under the appropriate conditions, all metals may experience this form of corrosion, such as general rusting of steel in a relatively uniform manner. When references are made to a corrosion rate, for example, if a reference is made to steel in a mild acid as estimated to corrode at 1.27 mm (50 mils) per year, general corrosion is the corrosion mechanism. The probable life of a component could thus be forecast with some accuracy. Unfortunately, corrosion is often localized corrosion, and general corrosion rates do not hold for localized corrosion.

Localized corrosion

Localized corrosion is a selective attack by corrosion at a small area or zone. It may be filiform, crevice, pitting or biological:[11] each has the common feature of being localized corrosion damage only.

An example would be the general appearance of a metal surface being

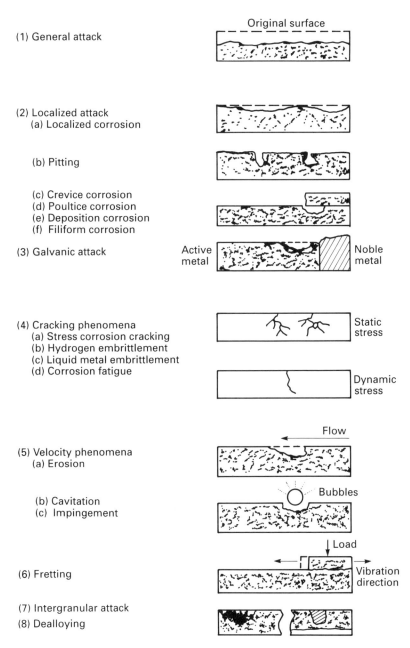

(1) General attack

(2) Localized attack
   (a) Localized corrosion

   (b) Pitting

   (c) Crevice corrosion
   (d) Poultice corrosion
   (e) Deposition corrosion
   (f) Filiform corrosion

(3) Galvanic attack

(4) Cracking phenomena
   (a) Stress corrosion cracking
   (b) Hydrogen embrittlement
   (c) Liquid metal embrittlement
   (d) Corrosion fatigue

(5) Velocity phenomena
   (a) Erosion

   (b) Cavitation
   (c) Impingement

(6) Fretting

(7) Intergranular attack

(8) Dealloying

4.3 Corrosion schematics (source: NACE, 1984, *Corrosion Basics: An Introduction*, NACE, Houston, TX, p. 94, reproduced with permission).

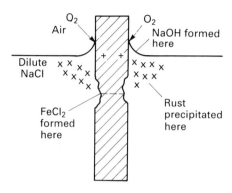

4.4   Splash zone. Differential aeration cell at waterline (data from Uhlig, H.H., and Revie, R.W., *Corrosion and Corrosion Control*, copyright © 1985 John Wiley & Sons. Reprinted by permission of John Wiley & Sons, Inc.).

adequate, but one region, such as at the 'splash zone' at a metal-to-water interface, suffering severe attack, as shown in Fig. 4.4. Commonly, large parts of the original surface are not attacked, or are attacked to a much lesser degree.

Pitting is the most common type of localized corrosion, whereby small areas of metal are dissolved by the corrosion process to produce pits. Pitting corrosion can be caused by a number of corrosion mechanisms.

Crevice corrosion

Crevice corrosion is a particular form of pitting corrosion that occurs between objects in contact with each other. Crevice corrosion typically occurs in cracks or crevices formed between mating surfaces. Crevices may be naturally occurring or synthetic.[12] It is commonly found, for example, at flanged connections. The two surfaces may both be metal or one may be nonmetallic, such as under a nonmetallic gasket. An example is shown in Fig. 4.5.

Crevice corrosion often initiates as a result of a differential aeration

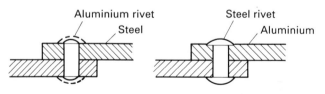

4.5   Galvanic couple between steel and aluminium.

Table 4.2. Factors that affect crevice
corrosion resistance[14]

Geometrical
Type of crevice
   metal-to-metal
   metal-to-nonmetal
Gap size
Crevice depth
Surface–area ratio

Environmental
Bulk solution
   O_2 content
   pH
   aggressive ion content
   temperature
   agitation
Mass transport
Diffusion
Biological

Electrochemical reactions
Metal ions
Oxygen reduction
Hydrogen evolution

Metallurgical
Alloy composition
Surface film characteristics

mechanism.[13] Oxygen in the liquid that is deep in the crevice is consumed by the corrosion reaction. The oxygen content of the liquid at the mouth of the crevice, which is exposed to air, is greater, creating a localized cell. The area of the metal being attacked is the anode, the surface in contact with the oxygen-depleted liquid. The pH changes at anodic and cathodic sites, which further stimulates the cell.

When the bulk environment is very aggressive, the general corrosion may preclude localized corrosion at a crevice.[14]

Crevice corrosion commonly occurs on film-protected (passive) metals, such as the austenitic stainless steels. Susceptibility to crevice corrosion can be reduced by good design that minimizes crevices, allows for adequate draining, and reduces the buildup of debris. Factors that affect crevice corrosion are shown in Table 4.2.

Galvanic corrosion

Galvanic corrosion generally occurs as a result of the potential differences between two metal surfaces, often two different metals, that are in contact with each other. Figure 4.2 is a typical flashlight battery, which illustrates a characteristic of galvanic corrosion to generate electricity.

A small electrochemical cell forms when there are two metals connected to each other in the presence of an electrolyte. This cell results in a current flow. The metal that is the anode will corrode preferentially to the metal that is the cathode. For example, in the flashlight battery the zinc is the anode and is electrically connected to the graphite, or carbon rod, which is the cathode. When they are in the presence of an electrolyte, current will flow between the two electrodes, causing corrosion of the zinc and producing a cathodic reaction at the graphite. The battery stops producing power when the zinc is depleted and completely corroded.[7]

Similarly, galvanic corrosion occurs when one metal is in contact with another metal and the two are in a conductive solution, thus producing a galvanic couple. The corrosion of a more active metal such as aluminium will be accelerated, while the corrosion of the other more noble metal, such as steel, will be reduced.[15] Which one will act as the anode and which as the cathode is determined by the direction the electric current flows as a result of the galvanic effect.[15]

The most direct way to determine current direction is by placing two metals in solution, providing an electrical path, and measuring the current flow. Arranging this information would produce a galvanic series, which could be repeated for many metals in many solutions. If the experiments were repeated for different solutions, values and the order of the relationships would change.

One commonly used galvanic series is for seawater. Table 4.3 shows the galvanic series in seawater. Note how an anode in one situation may be a cathode in another.

For an example of how the galvanic series can be used, consider a stainless steel nut and bolt attached to a carbon steel frame immersed in seawater. There is an electrochemical cell made up of the steel (anode), the stainless steel (cathode), the threaded connection (making a circuit) and the seawater (the electrolyte). Corrosion will occur at the anode. The amount of corrosion and the rate will depend on several factors, such as the size and geometry of the items.

The magnitude of the galvanic effects is a separate issue. The previous discussion concerned the direction of galvanic action as determined by the relative potentials of the two metals in a galvanic couple.[16] The intensity of the corrosion is determined by the amount of current, current density or current-per-unit-area, which refers back to Ohm's law, in which the amount of current that can flow is proportional to the voltage for any given value of resistance. When the potential difference between the galvanic couples is high, for example, zinc and copper at 700 millivolts in seawater, more current and therefore more corrosion is generated.[16]

Table 4.3. Galvanic series of commercial metals and alloys in seawater

Active or anodic (−)	Magnesium and magnesium alloys
	Zinc
	Galvanized steel
	Commercially pure aluminium
	Mild steel
	Cast iron
	Type 410 stainless steel (active)
	Type 304 stainless steel (active)
	Lead
	Tin
	Muntz metal
	Naval brass
	Nickel (active)
	Yellow brass
	Admiralty brass
	Copper
	70/30 copper nickel
	Nickel (passive)
	Type 410 (passive)
	Titanium
	Type 304 stainless steel (passive)
	Silver
	Graphite
Noble or cathodic (+)	Gold
	Platinum

Cracking phenomena

Cracking phenomena include several specific failure mechanisms:[17]

- Stress corrosion cracking (SCC), an anodic process.
- Hydrogen induced cracking (HIC), a cathodic process.
- Liquid metal embrittlement (LMC), a physiochemical process.

All three phenomena are sometimes referred to as SCC, or as corrosion accelerated by tensile stress. SCC, also called environmental cracking, is defined as 'a spontaneous brittle fracture of a susceptible material, usually ductile, under tensile stress in a specific environment over a period of time.'[17] Most commercial metals may crack in a brittle manner in specific environments. Details about these forms of cracking are discussed more fully in the NACE, 1984, *Corrosion Basics, An Introduction* and the ASM Metals Handbook, 1987, Volume 13, *Corrosion*.

Environmental cracks are microscopic when they originate; as the cracking proceeds, failure often occurs by overload. The cracking mode is either intergranular, transgranular or combinations of both. Intergranular means that the cracking is along the grain boundaries, while transgranular is across the grains. The mode of cracking may change, because of a change in temperature or pH, for example. Compressive stresses do not cause cracking. In fact, forcing the metal surface into compressive stress by shot peening reduces the possibility of fatigue or stress corrosion cracking.

Residual stresses may result from fabrication operations such as drawing, rolling or welding, and may be sufficient to produce SCC. In addition, cooling from a high temperature often induces residual stresses.

Corrosion can build up between surfaces and generated stresses. If the metal is in an environment/alloy combination susceptible to SCC, the stresses produced are often sufficient to cause SCC. Notches or stress risers often raise the apparent stresses in a localized area to a value high enough to induce SCC.

SCC is frequently an anodic process so applying cathodic protection prevents SCC.[18] The process starts with an incubation period, often termed a nucleation stage, when cracking begins at a microscopic level, followed by the cracks propagating. They often grow and then self-arrest. Typically, the cracking occurs in mildly corrosive environments. Note that if the stresses or corrosives are not present, cracking does not occur. All three criteria are needed for SCC to occur: a susceptible environment, a susceptible material and stress. Often, very little metal loss is associated with SCC, which makes it particularly unpredictable in terms of catastrophic failures.

Many metals are susceptible to SCC. The most common form of SCC in steel results when low and medium carbon steels are used for the transportation and storage of ammonia or nitrates. Steels also fail by SCC in a variety of environments, such as in mixtures of carbon monoxide, carbon dioxide and water and steam vapour. The presence of carbonates, and bicarbonates in certain conditions, as well as hot caustic solutions, also contributes to failures by SCC of steels.

Other examples of failures by SCC include aluminium and austenitic stainless steels in chloride environments. Many problems occur in the 18–8 grades (Types 304 and 316) of austenitic stainless steels. They are highly susceptible to SCC by chloride ions at temperatures greater than 60 °C (140 °F).[19] SCC is a common source of failures in stainless steel heat-exchanger tubes in chloride-bearing cooling waters.[19] Other deposits on the tubes aggravate the problem by trapping the chlorides and increasing their concentration. 'Super' stainless steels, such as the 6% molybdenum steels, were specifically developed to resist chloride SCC.

Titanium alloys are generally immune to chloride SCC, but do fail in certain environments. Particular conditions include high temperatures (400 °C; 750 °F) and highly stressed materials in the presence of sea salt, pure sodium chloride or concentrated chlorides.[19]

Hydrogen damage is another important form of cracking. Hydrogen produced by whatever means penetrates the metal as atomic hydrogen and can cause damage by hydrogen embrittlement or hydrogen blistering.[20] Hydrogen embrittlement is the loss of ductility of a metal because of the diffusion of hydrogen into the metal: the metal becomes brittle and fails. Often the hydrogen is trapped in pits, cracks or crevices; it concentrates and the metal is subject to failure at these locations. Ductility can be restored by heating the metal to force out the hydrogen.

Hydrogen blistering is similar. When atomic hydrogen diffuses into a metal, concentrates as a cavity or flaw, and forms molecular hydrogen, the gas produced can no longer migrate out of the metal by diffusion. This gas generates a high pressure and expands, producing a blister.

Velocity effects

Velocity effects include erosion–corrosion, cavitation and fretting. Erosion–corrosion is corrosion attack accelerated by high velocity flow or impingement by particles. If the flow of liquid over a metal surface becomes turbulent, the random liquid motion impinges on the surface to remove the oxide film, and erosion–corrosion occurs.

Cavitation is a specific form of velocity-effect corrosion caused by the implosion of bubbles. These bubbles form when the local pressure in the liquid flowing past the metal surface decreases below the vapour pressure. Cavitation damage is a form of localized corrosion combined with mechanical damage that commonly occurs in turbulent liquids and usually takes the form of pitted or rough surfaces.

Fretting is a form of velocity-effect corrosion in which motion of mating surfaces occurs under local load and vibration. It often takes the form of deep pits or localized surface discolorations.

Corrosion fatigue is not considered a velocity effect, but is often discussed in association with damage from combined mechanical and chemical effects. Corrosion fatigue is the tendency for a metal to fracture under repeated cycles of tensile and compressive stresses.[21] In a combination of stresses in a corrosive environment, the chemical and mechanical attacks reduce the corrosion resistance; failure may occur at lower than expected stresses.

Intergranular corrosion

Intergranular corrosion occurs on the microscopic level. A small amount of metal is preferentially attacked along grain boundaries. This action leaves the structure unstable and allows grains literally to fall out of the metal structure. It usually initiates on the surface and proceeds by local action in the vicinity of the grain boundary.[22]

The driving force for integranular corrosion is the difference in potential between the grain boundary and the bulk grains.[22] This difference may be caused by variations in the chemical composition between the areas. There may be a concentration of impurities or alloying elements at the grain boundaries and a second phase or constituent may form at the grain boundaries. This region may have a different corrosion potential from that of the grain, and thus will create a localized corrosion cell.

The grain boundary constituent may be anodic, cathodic or neutral to the base metal or adjacent area. This grain boundary area will corrode preferentially if the base metal is anodic to the other area.[22] When cracks develop, they will follow the grain boundary, breaking away the grains on the outer surface of the metal.

Intergranular corrosion reduces the elongation of a metal before any real loss of tensile strength occurs, creating the potential for unexpected catastrophic failure. Susceptibility to intergranular corrosion increases with heat treatment, because precipitation of second phase particles in the grain boundaries often forms. Cold work can also cause precipitation, and contribute to intergranular corrosion.

In some environments, such as after sensitization by welding has occurred, austenitic stainless steels are susceptible to intergranular corrosion. Sensitization, discussed in Section 3.6, is the susceptibility of stainless steels to intergranular corrosion. The sensitization occurs after thermal processing, which is often associated with welding. For austenitic stainless steels, the heat of welding causes a precipitation of chromium carbides around the grain boundaries. The region that is depleted in chromium carbide is less corrosion-resistant. As corrosion attack occurs, the grain boundaries are selectively attacked and the grains may eventually fall out.

Dealloying corrosion

Dealloying is the selective removal of a constituent of an alloy, a good example being the dezincification of brass. In this case, the zinc is selectively dissolved, and a porous structure results. The structure may seem to be intact, but in fact its mechanical strength is greatly compromised.

High temperature corrosion

The attack of metals at elevated temperatures is usually considered a form of oxidation. When a metal is oxidized at elevated temperatures, a stable oxide or similar compound coats the surface. The corrosion resistance at high temperatures is governed by one of two branches of physics, thermodynamics or kinetics.[23]

It is important to repeat that thermodynamics will tell us what *can* happen, while kinetics will tell us what *will* happen.[2] Thermodynamics deals with the conversion of heat into energy. Kinetics deals with the study of reaction rates and the factors that affect the rates.

4.3.2   Corrosion rate

The rate of corrosion depends on several factors.[16] Some of these factors include:

- Area effect.
- Polarization.
- Concentration cell effect.
- Galvanic effects.

The rate of corrosion is directly proportional to the current flow, as discussed above about Ohm's and Faraday's laws.

Area effect

Effects of current flow on corrosion are related not only to the total amount of current flow, but also to the current density; i.e. the current flow per unit area.[16] Thus, current over a small surface area has a more concentrated impact than current over a large area. The situation most favourable to corrosion production would be a very large cathode and a very small anode. For example, if large copper plates were connected with steel rivets and the component were then immersed in seawater, the steel would rapidly corrode.

Polarization

Polarization is the shift in electrode potential resulting from the effects of current flow measured against the zero-flow potential. That shift in potential change is caused by the net current towards or away from the electrode (measured in volts). Polarization occurring at the anode and cathode determines the corrosion rate.

The causes of polarization are concentration polarization, activation polarization (discussed above) and *IR* drop (defined below). Since there are two types of polarization, there are also two ways in which electrochemical reactions can be slowed down.

IR drop is the voltage that results from current flowing through a conductor. The difference between the starting potential and the final potential is the *IR* drop.

Why is polarization important? The higher the resistances, the lower the corrosion current. This concept of polarization is the fundamental feature of corrosion control using cathodic protection. Cathodic protection reduces corrosion by making the metal a cathode through an impressed current or a galvanic anode, so transfering corrosion from a structure to be protected to a ground bed. Polarization diagrams and the *IR* drop are discussed in Section 4.4.2.

Concentration cell effect

A concentration cell refers to an electrochemical reaction as a result of concentration changes in the solution adjacent to the metal surface. Corrosion cells are formed because of differences in the environment, and occur in two types: oxygen and metal ions.

Oxygen can maintain and promote a cathodic corrosion reaction in an aqueous environment, because the concentration of dissolved oxygen from one point on a metal surface to another varies. In a similar manner, metal ions may concentrate, driving the cathodic reaction, which occurs when the concentration of metal ions is higher at one point on a metal surface than another.

Galvanic effects

Galvanic action is controlled in part by the potentials of the metals in a galvanic couple.[16] The intensity of corrosion is determined by the current density, i.e. the current per unit area, and from Ohm's law, the amount of current that flows is directly proportional to the voltage for a given resistance. An example would be zinc and copper in seawater that generates a potential of about 700 millivolts. A galvanic couple of naval brass and copper generates about 40 millivolts. The zinc and copper galvanic cell generates more current, and therefore produces more corrosion. The flow of current between the two metals is sometimes referred to as the open circuit potential (OCP).

4.3.3 Reactions: oxidation and reduction

When corrosion occurs, the metal changes from the atomic to the ionic state. Two half reactions occur simultaneously:

- The oxidation reaction occurs when electrons are lost from an atom or compound.
- The reduction reaction occurs when electrons are gained from an atom or compound.

Oxidation occurs at the anode. Corrosion (the metal loss portion of the half reaction) is an anodic reaction, which is shown as:

$$M^0 \rightarrow M^+ + ne^- \qquad\qquad [4.16]$$

or, to give a specific example:

$$Fe^0 \rightarrow Fe^{2+} + 2e^- \qquad\qquad [4.17]$$

Reduction occurs at the cathode. This reaction is shown by the cathodic reaction, which is a reduction process:

$$H^+ + e^- \rightarrow H^0 \qquad\qquad [4.18]$$

In this reaction, the reduction of hydrogen, the hydrogen ions react with electrons to form atomic hydrogen.

Another important example is that of hydrogen ions and oxygen forming water:

$$O_2 + 4H^+ + 4e^- \rightarrow 2H_2O \qquad\qquad [4.19]$$

Equation 4.19 shows oxygen reduction, and what occurs in acid solutions. Oxygen reduction may also occur, as shown in Equation 4.20, in neutral or alkaline solutions:

$$O_2 + 2H_2O + 4e^- \rightarrow 4OH^- \qquad\qquad [4.20]$$

The anodic and cathodic reactions occur simultaneously and at the same rate.

pH and corrosion

The pH is the negative of the logarithm of the hydrogen ion concentration:

$$pH = -\log(H^+) \qquad\qquad [4.21]$$

When there is an excess of H^+ ions, the solution is acidic. When there is an excess of OH^- (hydroxyl) ions, the solution is alkaline (at room temperature). A pH of 7 is neutral; pH of lower than 7 approaches acidic, and of greater than 7 approaches alkaline. A pH of 1 indicates a very strong acid, while a pH of 14 indicates a very strong base, or alkaline.

The effect of pH on corrosion is complex. The importance of the hydrogen ion lies in its ability to interact with the metal surface. Many alloys, such as the austenitic stainless steels, form an oxidized surface film which often contains hydroxide-like species tied to water. One possible reaction is:[24, 25]

$$H_2O \Leftrightarrow OH_{adsorbed} + H^+ + e^- \qquad\qquad [4.22]$$

A glance at this equation shows that hydrogen ions, hydroxyl ions, electrons and water coexist on the metal surface. The interaction results in a rate equation[25]

$$r = kC_{H^+}{}^n \qquad\qquad [4.23]$$

where r is the corrosion rate, k is the rate constant, n is an exponent and C_{H+} is the hydrogen ion concentration. The influence of biological variables and local environments on rate equations can be is significant.

Temperature and corrosion

Temperature is another complex variable on corrosion. Temperature difference creates a heat flow that is a form of energy.[25]

One way in which temperature can affect corrosion is governed by the influence of rate as described by the Arrhenius rate equation:

$$r = A \exp(-E/RT) \tag{4.24}$$

where r is the corrosion rate, A is a pre-exponential factor, E is the activation energy, R is the gas constant and T is absolute temperature.

4.4   Fundamentals of electrochemistry

Electrochemistry is defined as a chemical science and technology concerned with the properties of ions in solution, and their reactions at metal–solution interfaces.[2]

While the study of electrochemistry is important for corrosion applications, it also covers other industries and disciplines. These include energy production and storage, electrolysis, extractive metallurgy, metal finishing and electroplating, water purification, medicine, storage batteries and many others. The key factor in all facets of electrochemistry is chemistry associated with electrical phenomena.[2]

4.4.1   Electrochemical reactions: half-cells

An electrochemical reaction is one in which a chemical reaction involves the transfer of electrons. It involves oxidation (anodic reactions) and reduction (cathodic reactions). To review, when electrons are lost from an atom or compound, an oxidation reaction has occurred. When electrons are added to an atom or compound, a reduction reaction has occurred.

In relating the potential of different metals to each other, the half-cell is the standard. Because absolute standards are unknown, we assume that the standard for the reaction

$$2H^+ + 2e^- \rightarrow H_2 \tag{4.25}$$

is equal to zero at all temperatures, giving a relationship with all values of electrode potentials with reference to the hydrogen electrode as

$$\phi_{H_2} = 0 - (RT)/(2F)\ln p_{H_2}/(H^+)^2 \tag{4.26}$$

Table 4.4. Potential values of reference electrodes[28]

Half-cell	Potential (V)
Saturated calomel	+0.2415
Normal calomel	+0.2800
Tenth normal calomel	+0.3337
Silver–silver chloride	+0.2222
Saturated copper–copper sulphate	+0.3160
Hydrogen	0.0000

where p_{H_2} is the fugacity of hydrogen in atmospheres and (H^+) is the activity of hydrogen ions.[26] Measuring the EMF of a cell, for example, of a zinc and hydrogen electrode in a zinc salt solution of known activity of Zn^{2+} ions and H^+ ions, the standard potential $\phi°$ for zinc can be calculated and found to be -0.763 V. This basis is termed the standard hydrogen electrode (SHE).

Current flows from the anode to the cathode in solution. Unfortunately, two sign conventions are used for the potential. This text assumes a metal such as zinc to have a negative potential, and a metal such as gold to have a positive potential relative to the hydrogen half-cell, which is zero. It is important to recognize the sign and the magnitude of the voltage.[27]

For convenience, other half-cells are frequently used. Table 4.4 shows their potentials relative to the standard hydrogen electrode. We often need to convert from one scale to another. For example, we usually use a saturated copper–copper sulphate electrode ($Cu/CuSO_4$) as the reference to measure the potential of steel pipe buried in soil. We would obtain a reading of -0.700 V, for example. To convert to the SHE scale, we add $+0.316$ V to get a total of -0.384 V versus SHE scale.

4.4.2   Polarization diagrams

Polarization is the flow of potential because of current flow. Figure 4.6 is a polarization diagram. The axes are E, potential, and $\log I$, current. Figure 4.6 shows the cathode potential and the cathodic polarization, as well as the anode potential and the anodic polarization. The point where the two lines intersect is E_{corr} and I_{corr}. E_{corr} is the free corrosion potential, also called the OCP. I_{corr} is the current density (current per unit area) at E_{corr} (or at the free corrosion potential).

This concept of polarization is partly based on the work of Tafel in 1905.[3] He found that the current, I, equivalent to the rate of a single electrode reaction, is related to the potential of the metal by the equation:

$$E = a + b \log I \qquad [4.27]$$

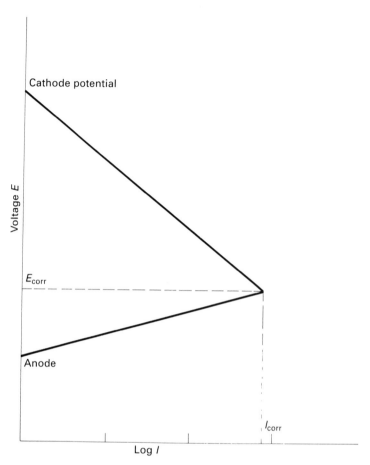

4.6 Polarization diagram.

The x-axis in the polarization diagram is $\log I$ and the y-axis is E (potential). The value $\log I$ is current density, or current per unit area. The value E_{corr} (OCP) is the potential when no current is flowing through the cell (made up of an anode, a cathode, an electrolyte and a circuit). The point E_{corr} is where the anode and the cathode potentials intersect.

Figure 4.6 represents the two reactions, anodic and cathodic, drifting toward each other until they intersect, as the corrosion reaction progresses. This intersection is E_{corr} and I_{corr}. Note that the rate of corrosion is directly proportional to current flow and that $E_{cell} = E_{cathode} - E_{anode}$.

What happens as factors affect polarization? We will examine the effects of several factors including current, surface area, temperature, agitation and concentration. By understanding what happens to the polarization, we can estimate the corrosion rate.

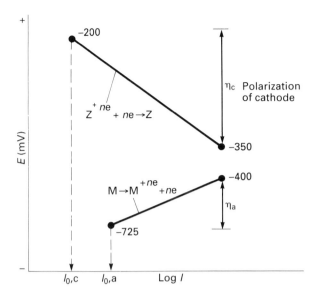

4.7 Anodic and cathodic activation polarization of corrosion cell.

Current flow

Figure 4.7 shows that the direction of the polarization shift is determined by the direction of the current flow. A positive shift occurs when ions in the electrolyte accumulate during the anodic reaction. Reduction of hydrogen reaction causes a negative shift, because it depletes the area of H^+ and increases H and H_2. A shift in potential of -0.200 V to -0.350 V versus $Cu/CuSO_4$ electrode would indicate a cathodic reaction. A shift in potential of -0.700 V to -0.400 V would indicate an anodic reaction.

Surface area

As Fig. 4.8 illustrates, the direction of polarization shift is affected by the surface area of the electrode. If the surface area on the cathode is large, the anode polarization is high and the cathode polarization is low. If the anode area is large, the anode polarization is low and the cathode polarization is high. Current concentrated on a small area will produce a greater effect than it would over a large area.

Temperature

Figure 4.9 shows how temperature affects the kinetics of both the anodic and cathodic reactions. Increasing temperature increases the current and

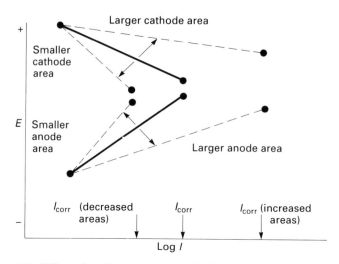

4.8   Effect of surface area on polarization.

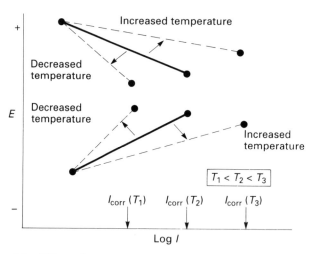

4.9   Effect of temperature on polarization.

decreasing temperature decreases the current. In addition, the diffusion rate is increased by increasing the temperature.

Agitation

Figure 4.10 shows how agitation affects the polarization. When there is little movement in a stagnant system, for example, polarization increases.

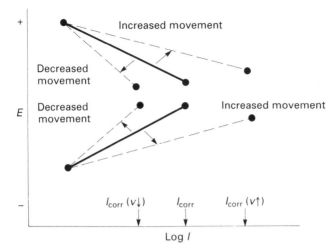

4.10   Effect of relative electrolyte/electrode movement on polarization.

For an agitated system, the reactants are replenished at the anodes and cathodes, and the polarization decreases.

Concentration

Figure 4.11 shows how concentration shifts the polarization. Increased metal ion concentration increases polarization of the anode, decreased metal ion concentration decreases the polarization of the anode.

4.4.3   Passivity and protective films

Passivity is a condition, usually one in which an active metal becomes more cathode-like, and often the result of a protective oxide film on the surface of the metal. This film is usually a few angstroms thick.

The film is sometimes a temporary condition or very fragile. Such is the case for some irons and steels in certain environments. The film may be an inherent property of the metal, as it is for gold and platinum (these are noble metals which are very resistant to most oxidation reactions). Alternatively, the protective film may be an oxide, as is the case for stainless steel.

4.5   Passivity and passive films on stainless steels

Figure 4.12(a) shows a typical anodic polarization curve for alloys that exhibit active–passive behaviour, such as the 300 series stainless steels. A

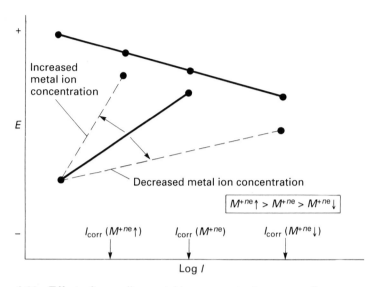

4.11   Effect of corroding metal ion concentration on anodic polarization.

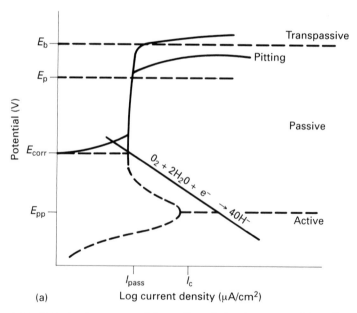

4.12   Polarization curves: (a) anodic polarization curve; (b) cyclic anodic polarization curve; (c) potential versus log I plot.

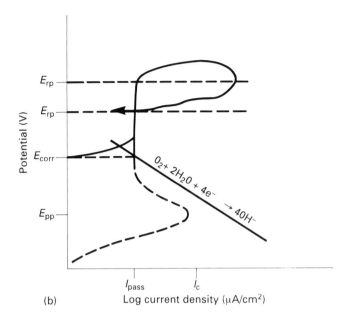

(b)

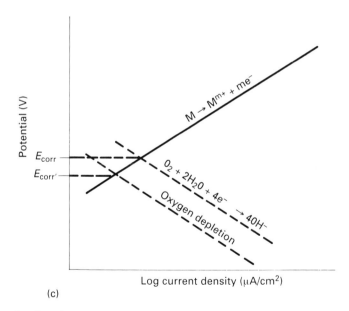

(c)

4.12 *Continued*

cathodic curve for the reduction of dissolved oxygen is also shown. The complete anodic and cathodic responses are shown as dashed curves. The as-measured anodic curve (i.e. E versus log of external current density) is shown as a solid curve.

The behaviour of the metal divides into three regions: active, passive and transpassive. In the active region, the behaviour of the metal is such that an increase in the applied potential causes a correspondingly rapid increase in corrosion rate. When the applied potential is increased sufficiently, the corrosion rate suddenly decreases. This behaviour corresponds to the beginning of the passive region. Further increases in applied potential produce little change in the anodic dissolution rate until eventually, at a very high applied potential, the corrosion rate begins to increase with increasing potentials in a region called the transpassive region, where the protective passive film is thermodynamically unstable.

Included in the plot of Fig. 4.12(a) are various terms used to define the behaviour of the alloy. A variable of particular importance for passive alloys is termed the pitting potential (E_p), which denotes the potential above which pits may nucleate and grow, but below which pitting is unlikely to occur.[29]

The good corrosion resistance that austenitic stainless steel alloys exhibit in an oxidizing environment begins with passivity. The passive film consists of a thin protective oxide film. The anodic polarization curve, shown in Fig. 4.12(a), has a typical S-shape for alloys that obtain their corrosion resistance from passive film formation. The shape of the anodic polarization curve is a function of the composition, thickness and properties of the film as it is formed upon passing from the active to the passive region.

Additional terms included in the plot of Fig. 4.12(a) and in the cyclic anodic polarization curve of Fig. 4.12(b) aid in defining the behaviour of the alloy. These terms include critical current density (I_c), passive current density (I_{pass}) and the primary passivation potential (E_{pp}). The term pitting potential, E_p, is the potential above which pits nucleate and develop. The term repassivation potential, E_{rp}, is the term for the potential where the reverse scan current density equals the forward scan current density, and generally represents the potential at which active pits repassivate.

To clarify the distinction between E_p and E_b, shown in Fig. 4.12(a), E_b occurs as the result of alloying the metal with chromium. For instance, pure chromium in chloride-free sulphuric acid solution exhibits a wide passive range, as shown in Fig. 4.12(a). The extreme of the passive range is determined by the onset of transpassive behaviour (E_b). The introduction of chloride ions lowers the noble extreme of the passive range to the pitting potential, E_p.

An alloy with good corrosion resistance will produce a curve with low values of I_c, I_{pass} and E_{pp}, and a high value of E_p. The location of this potential, E_p, depends on alloy composition, pH and other variables such as chloride concentration.

Open cell potential (OCP or E_{corr}) is the potential of an electrode measured with respect to a reference electrode or another electrode when no current flows to or from it. In terms of polarization behaviour, Fig. 4.12(c) is a curve showing active anode behaviour. On a plot of current density versus potential, two reactions are proceeding simultaneously. The anodic reaction is Equation 4.3, metal going to metal ions. The cathodic reaction is the reduction of oxygen. An approximation of E_{corr} is the intersection of these two curves. If the oxygen decreases, the curve drops; there is a different point of intersection, and E_{corr} occurs at a lower potential, also denoted as E_{corr}', in Fig. 4.12(c).

Another often used electrochemical test is the galvanic couple test. When dissimilar metals in electrical contact with each other are exposed to an electrolyte, a current, called a galvanic current, flows from one to the other.[23] This test aids in predicting or controlling how much galvanic couple action occurs when dissimilar metals are in contact under certain conditions.[31]

4.5.1   *The electrochemical behaviour of active–passive metals*

According to the oxide film theory, the structures of the film on stainless steels are complex. In addition, these thin films, 10–50 angstroms thick, consist of complex chemical compositions.[32] The films are generally hydrous oxides, enriched primarily with chromium, silicon and molybdenum.[33]

Okomoto presented a model for the film on stainless steel, as indicated in Fig. 4.13.[34] The film contains a gel-like structure. A metal ion is absorbed or replaced by a water molecule that changes the structure. This change, illustrated in Fig. 4.13(a) and (b), shows how ions are captured to form the film. In slightly more detail, hydrogen (H^+) is consumed and simultaneously a metal ion (MOH^+) is surrounded by water and precipitates as a solid film. With time, the film loses the water and changes to the less hydrated structure. At different stages of ageing, the film contains different bridges between the metal ions. The three bridges, H_2O-M-H_2O, $-HO-M-HO-$ and $-O-M-O-$, exist depending on the degree of loss of protons. The undeveloped (H_2O-M-H_2O) part of the structure reacts the most to corrosion.

The anodic polarization curve (Fig. 4.12) illustrates the corrosion resistance of austenitic stainless steels. This response to current and potential illustrates a typical active–passive behaviour for stainless steel. As discussed earlier, the overall shape of the curve indicates the corrosion behaviour in the test solution.[34] We can assess the degree of passivation

4.13 Metal ions dissolved through the undeveloped part in the film: (a) ions are captured to form the film; (b) ions due to the bridging of OH bond to metal ions (reprinted from *Corrosion Science*, **13**, Okomoto, G., 'Passive film of 18-8 stainless steel structure and its function', 471–479. Copyright © 1973, with kind permission from Pergamon Press Ltd., Headington Hill Hall, Oxford OX3 0BW, UK).

and the stability of the passive film by observing the current in the passive region and the potential at the pitting transpassive region. Lower currents in the passive region indicate a higher degree of passivation. A pitting or transpassive region at a more positive potential indicates that the passive film has greater stability.[34]

4.5.2   The pitting behaviour of metals

In some electrolytes, generally those containing halogen ions such as chlorides, the stability of the passive film is reduced.[29] In a manner similar to that illustrated in Fig. 4.12, this reduction is illustrated in Fig. 4.14: as the potential increases, the chloride ion reduces the width of the passive range. The chloride ions penetrate and destroy the film, thus decreasing its stability.[29]

After a pit initiates, it may grow or repassivate. The initiation stage usually requires a long time; however, after the pit initiates, the propagation rate can be rapid. In an aerated, neutral solution, the initial oxidation and reduction reactions are:

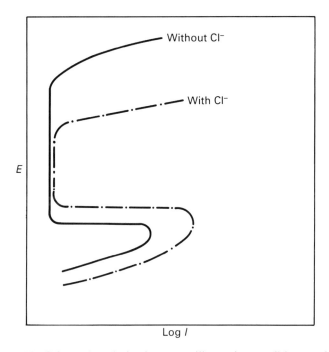

4.14 Schematic polarization curve illustrating conditions under which pitting may or may not occur (source: Peckner, D., and Bernstein, I.M., *The Handbook of Stainless Steels*, © 1977 McGraw-Hill, Mew York, p. 16.4, reprinted with permission).

$$M \rightarrow M^+ + e^- \tag{4.28}$$

$$O_2 + 2H_2O + 4e^- \rightarrow 4OH^- \tag{4.20}$$

When a pit initiates, the condition illustrated in Fig. 4.15 occurs. Initially, the above reactions occur uniformly over the whole surface. During the initial pit formation, possibly initiating because of a weak spot in the passive film, the oxygen within the pit is depleted by the reduction reaction. The restricted conditions in the pit do not allow replenishment of the oxygen. On the unpitted areas, the rate of corrosion remains the same as the passive current. However, to support the oxidation reaction within the pit, the rate of oxygen reduction increases on the passive surface and within the pit, an excess of metal ions (M^+) is produced, which is balanced by chloride ions, as shown in Fig. 4.15. The resulting reaction is:

$$MCl + H_2O \rightarrow MOH + H^+ + Cl^- \tag{4.29}$$

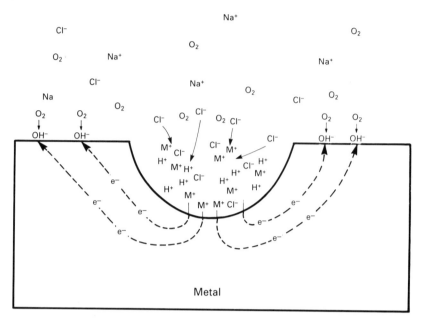

4.15   Autocatalytic processes occurring in a corroding pit (source: Fontana, M.G., and Greene, N.D., 1978, *Corrosion Engineering*, © 1978 McGraw-Hill, New York, p. 52, reproduced with permission).

The result is a high concentration of chloride ions in the pit and a correspondingly lower pH in comparison with the bulk solution. These effects further accelerate corrosion within the pit.

A narrow opening in a metal surface, at a joint between two surfaces or beneath a solid particle, provides conditions for an oxygen concentration cell known as crevice corrosion. In many ways, crevice corrosion is similar to a pit that has initiated. Similar to pitting, albeit another form of localized corrosion, the creviced region can become depleted in oxygen, undergoing acidification and chloride concentration; as a result, the passive film often becomes unstable. The passive film requires oxidizing conditions for stability, both to initiate passivity and to repair subsequent chemical or mechanical damage to the film. The oxygen-depleted conditions make crevices ideal sites for the existence of anaerobic microbes, and for MIC to occur. Conversely, biofilms can establish 'living crevices' that form on passive surfaces and produce deaerated conditions.[35]

According to Szklarska-Smialowska, the following features are thought to influence the formation of weak spots in the passive film, in terms of chemical inhomogeneities:[29]

- Boundaries between the metal matrix and nonmetallic inclusions, where differences in coefficients of thermal expansion exist.
- Boundaries between the metal matrix and second phase precipitates, which often have the ability to draw from the alloy some components responsible for the passive state (for example, depletion of chromium).
- Inclusions having greater chemical reactivity compared with that of the metal or alloy itself.

In terms of austenitic stainless steel, particular attention has been paid to the influence of sulphide and oxide inclusions. The role of pit nucleation sites influenced by the content and distribution of sulphides is important. Note that Poyet and Desestret showed in their studies of stainless steel that corrosion always initiates at nonmetallic inclusions, and the rate decreases with increasing alloy purity.[36] Lorenz and Medawar showed the susceptibility of Cr–Ni–Mo stainless steels to pitting in chloride solutions, which depends on the presence of segregations and nonmetallic inclusion.[29, 37] In addition, work by Forchhammer and Engell characterized MnS as the most effective pit nucleation sites.[38] They determined that the pit density depended on heat treatment. Annealing caused the sulphides to coalesce and resulted in fewer pits. Wilde and Armijo found an accelerating pitting effect of sulphur while studying stainless steels in 10% ferric chloride solution.[39] The pits initiated at sulphide inclusions and the intensity of the attack increased with increasing sulphur content. Szklarska-Smialowski studied pitting in austenitic stainless steel and concluded that oxide inclusions did not nucleate pits. The pits nucleated preferentially at the sulphide inclusions.[40, 41]

None of the above implies that pits nucleate at all sulphide inclusions. Rather, Szklarska-Smialowska found that nucleation seems to depend on pit geometry and chemical composition, where some inclusions are more likely to nucleate pits than others. Pits that have already formed may also protect against further pitting in the surrounding area, where there are less efficient inclusions.

The boundary between separate phases is also a favourable site for pit nucleation. Devine found that in as-annealed duplex Type 308 stainless steel, pits nucleated preferentially at the austenite/ferrite boundaries.[42] He concluded that this was due to the high concentration of sulphur at the interface, or a high sulphide density along the interface. Others have shown that the austenite/ferrite interfaces are contaminated with sulphur or sulphides.[29, 42, 43]

The term 'induction time' for pit nucleation describes the time required to form the first pit on a passive metal in a solution containing aggressive anions (for example, chloride ions). According to Szklarska-Smialowski, at a constant potential higher then the pitting potential, E_p, the induction times depend on the chloride concentrations.[29] Conversely, at a constant chloride concentration, the induction time is a function of the applied potential. For a given metal, the induction time decreases when the concentration of chloride ions or the potential increases; admittedly, the induction time depends on the quality of the passive film, as well as on the experimental conditions.

Experiments that determine the induction time as a function of film thickness clarify the mechanism of pit initiation. For example, the induction time is not always dependent on film thickness, as Okomoto showed.[34] He showed that film structure is more important than film thickness for pit initiation on 18Cr–8Ni stainless steel. He concluded that the amount of water bound in the film was responsible for such behaviour.

Golovina thought that pit formation is better described as the result of a competition between two processes:[44]

- Passivation caused by the adsorption of oxygen-containing molecules from the solution.
- Activation caused by the adsorption of aggressive anions. The current is then controlled by the rates of each process.

4.5.3   Crevice corrosion

There are both similarities and differences between pitting and crevice corrosion. Pitting initiates at susceptible points on a metal surface where the surrounding solution has free access.[29] Crevice corrosion limits its attack to restricted areas of the surface, where access to the environment is defined by geometry.

Crevice corrosion is accelerated by the stagnant solution in the gap. Pitting often initiates at discontinuities in the metal surface, such as nonmetallic inclusions, grain boundaries and other surface features.[29] Pitting and crevice corrosion are related, however, especially in the case of chloride-assisted corrosion. According to Szklarska-Smialowska, the critical potential at which crevice corrosion nucleates is usually less positive than the pitting potential. Crevice corrosion often initiates more rapidly than pitting, which partially explains why crevice corrosion is often more detrimental than pitting.

The mechanism of crevice corrosion was explained earlier. Assume that a metal with a crevice is immersed in a neutral chloride solution. The creviced region becomes depleted in oxygen and undergoes acidification and chloride concentration; as a result, a differential aeration cell forms. According to Szklarska-Smialowska, the critical acidity at which crevice corrosion initiates within a crevice is a function of several factors:[29]

- The composition of the alloy.
- The critical passivation current, depending on the composition of the bulk electrolyte.

Table 4.5. Environmental factors influencing corrosion[47]

- Temperature
- Pressure
- Velocity
- Impurities
- Mechanical effects
- Biological factors
- Stray currents

- The crevice geometry.

Four stages occur in the overall crevice corrosion process at open circuit potential in aqueous solutions containing chloride ions and dissolved oxygen:[29]

- Depletion of oxygen from the crevice electrolyte.
- An increase of acidity and Cl^- concentration in the crevice electrolyte.
- A permanent breakdown of the passive film and the onset of rapid corrosion.
- Crevice corrosion propagation.

4.6 The influence of system parameters on electrochemistry and corrosion

Often, corrosion control requires a change of metal or a change of environment.[45] We usually cannot completely control the environment in which a metal operates. In addition, the temperature, pressure and chemical composition of the system often change from the original design.

4.6.1 Environmental effects

Changing materials is often an expensive way to control corrosion. In addition, some specialty metals may be resistant to general corrosion but susceptible to localized corrosion or stress corrosion cracking, and thus unsuitable for the intended service.[46] Some environmental factors influencing corrosion are shown in Table 4.5.

4.6.2 Factors affecting corrosion

In designing materials specifically to reduce corrosion, metallurgists must consider the following factors.

The effect of temperature and pH on pitting

The temperature increase in a corrosive system increases the rate of corrosion, since the kinetics, or the rate of a reaction, have increased.[48] Such is the case even for components in liquid solutions that are near room temperature. The section with the higher temperature may be anodic to the other.

In addition, the ionization constant of water increases with temperature.[49] Pure water with a pH of 7 at a given temperature will therefore have a lower pH at a higher temperature. For example, increasing the temperature can change the water's pH from neutral to acidic, which in turn affects the corrosion.

The effects of alloying elements and heat treatment on pitting

In general, chromium and molybdenum increase the pitting resistance of steels in a chloride environment. Molybdenum additions to austenitic stainless steel alloys improve their passive properties and their resistance to pitting.[50] These alloying elements reduce the susceptibility to pit initiation and the rate of pit propagation. How other elements influence pitting is still being investigated.[51] Streicher studied the susceptibility of steels to determine the optimum alloying composition with the highest ductility and resistance to pitting in 10% ferric chloride solution. He found this alloy to be Fe–28Cr–4Mo, with carbon not to exceed 0.01% and nitrogen not to exceed 0.02%.

Brigham and Tozer studied the effect of molybdenum on pitting susceptibility for austenitic stainless steel.[52] They found a linear relationship between molybdenum content and critical pitting temperature.

The results distinguishing the effects of nitrogen in steel as an alloying element appear to conflict, which is probably because of the combined effects of other elements on the influence of nitrogen. In addition, the nitrogen can form nitrides or carbo-nitrides and precipitate as separate phases. These separate phases increase susceptibility to pitting corrosion.[53] Carbon in the form of precipitated carbides increases the susceptibility of austenitic stainless steel to pitting corrosion, particularly in the sensitized condition, as discussed earlier.

Table 4.6 summarizes studies done to improve the corrosion resistance of stainless steels. In general, increasing the chromium content beyond 12% yields steels that can be passivated in more aggressive electrolytes. Alloying with both nickel and chromium increases the ability to passivate.

In terms of heat treatment effects, as discussed earlier in this chapter, austenitic stainless steels heated in certain temperature ranges may undergo sensitization and be prone to intergranular corrosion. At the sensitizing temperature, the chromium carbides precipitate at the grain boundaries,

Table 4.6. The effects of alloying on the pitting resistance of stainless steel alloys[54]

Element	Effect on pitting resistance
Chromium	Increases
Nickel	Increases
Molybdenum	Increases
Silicon	Decreases, increase of Mo is present
Titanium and columbium	Decreases resistance in $FeCl_3$; other mediums no effect
Sulphur and selenium	Decreases
Carbon	Decreases, especially in sensitized condition
Nitrogen	Increases

which depletes the chromium of the adjacent metal. In general, sensitized steels are susceptible to intergranular corrosion and pitting.[55]

The resistance of passive films to breakdown is related to the particular base metal properties. If zones adjacent to the grain boundaries are impoverished in chromium, as occurs in the sensitized condition of stainless steels, the passive film is less protective in this region, and localized corrosion may occur. At sites where nonmetallic inclusions occur on the surface, the film is weaker and pits may nucleate.

The effects of composition of the electrolyte on pitting

Most electrochemical reactions proceed at a higher rate at higher temperatures.[39] For example, consider austenitic stainless steel in nitric acid. Increasing the temperature of nitric acid greatly increases its oxidizing power. At low or moderate temperatures, such an increase would place the steel in the passive or transpassive region on the anodic polarization curve, as illustrated in Fig. 4.12. An increase in oxidizing power causes a rapid increase in corrosion rate.[56]

Many investigators researching the effect of solution acidity on pitting of iron-based alloys report that the pitting potential, E_p, shows small changes within a large pH range.[57] In very acidic solutions, however, the iron-based alloys typically result in severe general corrosion, while in very alkaline solutions, weak corrosion or sometimes inhibition occurs.

Different anions can cause pitting corrosion. The most aggressive seem to be chloride, which produces pits in iron, nickel and many alloys. According to Szklarska-Smialowska, other anions, such as sulphate ions, for example, can break down the passive film and cause pitting, much as chloride does.[57] Clearly, the electrochemical processes associated with corrosion are strongly affected by the ions in the electrolyte.

The pitting of weldments in austenitic stainless steel

The weld metal is less resistant to pitting than the base metal for austenitic stainless steels, because of differences of composition and structure resulting from the heating and cooling rate. Weld metal properties may differ from the base metal even for the same composition, which is usually the result of the weld being inhomogeneous. The weld typically contains ferrite, which is not present in the base metal. Garner studied the pitting of austenitic stainless steels in both the welded and unwelded conditions.[58] He found that autogenous welding had a detrimental effect on pitting resistance. The pitting potential and the critical pitting temperature were lower for welded than unwelded steel.[58,59]

Stalder and Duquette found a higher pitting potential for welded 304L stainless steel with 7–10 FN than for unwelded fully austenitic stainless steel.[50] They found that for the duplex structure, pit initiation occurred in the austenite phase, and primarily on the ferrite–austenite boundary.

Other studies showed that pits can develop as a result of the action of a macrocell between the anodic weld metal and cathodic base metal, which is considered significant in terms of MIC.[51,52] This area of electrical activity may influence the bacteria to select weldments for colonization, thereby leading to localized corrosion cells and subsequent pitting corrosion.

4.6.3   Electrochemical and corrosion factors affecting MIC

Considering the effects of solution changes on microbiological processes from an electrochemical approach requires examining many factors. These include anodic and cathodic reaction rates, corrosion rates, solution modifications such as changing oxygen content and pH, and alloy compositions.[61]

The influence of anodic and cathodic reactions

Buchanan and Stansbury discuss the anodic and cathodic reactions as related to MIC.[61] For electrochemical corrosion, they developed a model in which the anodic reaction (oxidation or corrosion reaction) is generalized as:

Metal $\rightarrow$ corrosion product + electrons

The cathodic reactions are:

$$O_2 + 4H^+ + 4e^- \rightarrow 2H_2O \qquad \text{at pH} < 7$$
$$O_2 + 2H_2O + 4e^- \rightarrow 4OH^- \qquad \text{at pH} > 7$$
$$2H^+ + 2e^- \rightarrow H_2 \qquad \text{at pH} < 7$$
$$2H_2O + 2e^- \rightarrow H_2 + 4OH^- \qquad \text{at pH} > 7$$

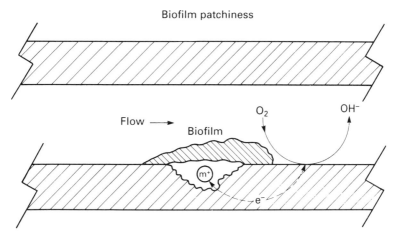

4.16   The physical presence of microbial cells on a metal surface, as well as their metabolic activities, impacts electrochemical processes. The adsorbed cells grow and reproduce, forming colonies that constitute physical anomalies on a metal surface resulting in formation of local cathodes or anodes (source: Characklis, W.G., Little, B.J., and McCauley, M.S., 1989, 'Biofilms and their effect on local chemistry', in *Microbial Corrosion: 1988 Workshop Proceedings*, EPRI EDR-6345, EPRI, Palo Alto, CA, reproduced with permission).

in which the first two reactions contain dissolved oxygen and the second two are deaerated. This model shows that the cathodic reactions are strong functions of dissolved oxygen concentration, and of pH. The most important part of the reaction occurs at the metal surface interface since the rates at which the reactions occur are a function of the potential at the interface. Polarization curves show this dependency with the $E \log I$ relationship, as discussed in Section 4.4.2. The corrosion potential, E_{corr}, and the corrosion current density, i_{corr}, are the OCP intersection of the two curves for the anodic and cathodic reactions.

The influence of solution modifications

Buchanan and Stansbury discuss modelling an interfacial solution undergoing continuous decreases in oxygen and pH owing to microbial activity.[61] The complexities of such modelling are apparent. The model assumes that the interfacial solution changes occur uniformly over the alloy surface. With MIC, such is not the case. The surface is composed of patchy regions of exopolymers and organisms attached in an unpredictable fashion (Fig. 4.16).[62] As a result, some regions are exposed to very aggressive environments, while others are exposed to the bulk solution.

Chlorides

Chlorides are particularly corrosive toward stainless steels. For example, in a study by Stoecker and Pope, a low-chloride water of 20 ppm chlorides in combination with MIC resulted in corrosion attack.[63] Depending on oxygen concentration and temperature, 20 ppm chlorides should be non-corrosive to stainless steel.[63] In standard pitting, chloride ions can be electrochemically transported to the anodic corrosion sites (pits) and accelerate the corrosion process, even when they are not the initiating factor.

Although this theory is controversial, many researchers accept the idea that iron bacteria of various types attack stainless steel, particularly at weld joints.[64] Some filamentous iron bacteria oxidize ferrous ions to insoluble ferric oxide/hydroxide, which forms a common sheath for several cells and produces a characteristic stalk-like form.[64] Jones suggests that the insoluble product is a mixture of ferric oxide (Fe_2O_3) and ferric hydroxide ($Fe(OH)_3$), in which an approximate formula is $FeO[OH]_n$.[64] The ferric ions are possibly not totally precipitated, especially in the acidic crevice regions. Ferric ions in solution serve as highly oxidizing species and tend to accelerate corrosion. Moreover, in waters containing chloride ions, iron-oxidizing bacteria may be directly involved in the production of ferric chloride, which is an extremely corrosive substance that can concentrate under nodules.[54, 64] Chloride ions can migrate into a crevice location by neutralizing the increased charge via anodic dissolution and then combine with the oxidation by bacteria of ferrous and manganous ions to ferric and manganic ions. A proposed mechanism for MIC attack is the result. Such an attack is sometimes termed MIC chloride-assisted attack.[65, 66]

4.6.4 Atmospheric corrosion

In general, corrosion rates are lower in air than in natural water or soil. The exception would be a corrosive atmosphere, in which specific agents accelerate corrosion. These include dust content, gases and moisture.[67]

Vernon performed experiments that show the importance of atmospheric dust.[68] His experiments exposed steel specimens to an indoor atmosphere. Some specimens were protected by a cage of muslin around the specimen. After several months, the unscreened specimens showed rust and weight gain while the screened specimens showed no rust and negligible weight gain.[67]

Average city air contains $2 \, mg/m^3$, with higher values for industrial atmospheres in the range of $1000 \, mg/m^3$ or more.[67] Uhlig estimates that more than 100 tons of dust per square mile (30 tonne/km^2) settle every month over an industrial city. A 1971 report estimated that 475 000 tons per month of particulates are produced by electric utilities and industrial coal-fired power plants in the United States.[69]

Sulphur dioxide, from the burning of coal, oil and gasoline, contributes to atmospheric corrosion problems. The sulphur content in fuels and associated problem chemical contents are being reduced in the United States, because of local and federal regulations.

Atmospheres are commonly divided into four types: industrial, marine, rural and indoor. Industrial atmospheres are characterized by pollution composed of sulphur compounds and nitrogen oxides.[70] Typically, sulphur compounds pick up moisture and form sulphurous acid, which combines with dust and settles on exposed surfaces, producing a wet corrosive film.

Marine environments are characterized by a fine mist composed of particles of sea salt that settle on exposed surfaces.[70] The distance from the ocean and wind is a critical factor in the extent of damage. The splash zone, the region that alternates between wet and dry and is subject to heavy spray and splashing, is considered intermittent immersion, and is one of the areas that most commonly suffers severe corrosion.

A rural atmosphere is usually characterized by organic and inorganic dust, instead of chemicals.[70] For example, moisture combines with carbon dioxide and forms a weak acid.

Finally, an indoor atmosphere may be the most varied and unpredictable, and is generally considered noncorrosive. In fact, an enclosed space may collect fumes if not properly ventilated, and, if moisture is present, could result in a highly corrosive environment.[70]

In general, weight loss by atmospheric corrosion of steels over time decreases as the length of time increases.[71] Often the equation

$$W = Kt^n \qquad [4.30]$$

is used, where W is the weight of the metal lost, t is time in years, and K and n are constants that depend on the alloy system and exposure site.[71] This equation use is limited for most applications, but can help in predicting long-term behaviour in some situations. Essentially, we cannot estimate the extent of corrosion with a single parameter, although most corrosion data try to do so. When the results of a several-year corrosion test are summarized in a single value, the kinetics of the system are lost, and results may be very unreliable.[72] See *Atmospheric Corrosion*, edited by W. Ailor, John Wiley & Sons, New York, 1982, for more detailed information.

4.6.5   Corrosion in soils

Underground corrosion and corrosion in soils is very important, given the existence of millions of kilometres of buried pipe, primarily for water, gas and oil.[73] A US 1978 report estimated a $158 million total cost for

Table 4.7. The relative values of soil corrosivity versus resistivity[75]

Resistivity (Ω cm)	Corrosivity
Under 500	Very corrosive
500–1000	Corrosive
1000–2000	Moderately corrosive
2000–10 000	Mildly corrosive
Over 10 000	Slightly corrosive

maintenance, repair and replacement costs due to corrosion of these structures in the United States.[74]

Corrosion in soil is similar to atmospheric corrosion, in that rates vary with the type of conditions present and the type of soil. When metals are buried in soils, specific conditions of soil composition, pH, moisture content and organisms result in widely varying performances.

In Section 4.3, we discussed resistivity and conductivity. To review, the electrical resistivity, ρ, is the electrical resistance offered by a material to the flow of current, multiplied by the cross-sectional area of current flow per unit length of current path. Technicians make electrical soil resistivity measurements, usually of soil or water, to obtain relative values for corrosion data for use in design. They typically use the ohm-centimetre as the unit of measurement and generally make these measurements before designing a cathodic protection system. Table 4.7 shows a rough estimation of values to use as a guide for determining soil corrosivity.[75]

If researchers test soil samples in the field, they typically do so using a soil resistivity measurement device, such as the ASTM G57 'Field measurement of soil resistivity using Wenner four-electrode method'. If researchers bring soil samples back from the field, they typically report them as conductivity or specific conductance in the units $\mu\Omega$/cm, or the equivalent unit μS/cm.

Carbon steel in soil results in corrosion, depending on the nature of the soil and other environmental factors, such as moisture and oxygen content.[71] Variables may affect the anodic and cathodic polarization of metals in soil.[76] Factors affecting the corrosivity in soil include:[77]

- Aeration.
- Electrical conductivity or resistivity.
- Dissolved salts.
- Moisture.
- Acidity or alkalinity.

Soils with high moisture content, high electrical conductivity, high acidity and high dissolved salts will be very corrosive.[72] The effect of aeration on soils is different from that on water. Poorly aerated wet soil may lead to increased attack by anaerobic bacteria, such as the sulphate-reducing bacteria.

See *Underground Corrosion*, Romanoff, M., National Bureau of Standards, Circ. 579, 1957, and *Underground Corrosion*, edited by Escalante, E., STP No. 741, the American Society for Testing and Materials, 1981.

Suggested reading

1   ASM Metals Handbook, 1985, *Desk Edition*, ASM International, Metals Park, OH.
2   ASM Metals Handbook, 1987, Vol. 13, *Corrosion*, ASM International, Metals Park, OH.
3   Fontana, M., 1986, *Corrosion Engineering*, McGraw-Hill, New York.
4   NACE, 1984, *Corrosion Basics, An Introduction*, Houston, TX.
5   Szklarska-Smialowska, Z., 1986, *Pitting Corrosion of Metals*, NACE, Houston, TX.
6   Baboian, R., ed., 1986, *Electrochemical Techniques for Corrosion Engineers*, NACE, Houston, TX.
7   NACE, 1982, *Forms of Corrosion Recognition and Prevention, Handbook 1*, Houston, TX.
8   Wulpi, D.J., 1966, *How Components Fail*, ASM International, Metals Park, OH.
9   Ailor, W.H., 1971, *Handbook on Corrosion Testing and Evaluation*, John Wiley & Sons, New York.
10 NACE, 1984, *Corrosion Data Survey*, 2 vols, Houston, TX.
11 Uhlig, H.H., 1948, *Corrosion Handbook*, John Wiley & Sons, New York.

References

1   Piron, D.L., 1991, *Electrochemistry of Corrosion*, National Association of Corrosion Engineers, Houston, TX.
2   Dillon, C.P., 1986, *Corrosion Control in the Chemical Process Industries*, McGraw-Hill, New York, p. 236.
3   Tafel, J., 1905, *J. Z. Physik. Chem.* **50** 641.
4   Szklarska-Smialowska, Z., 1986, *Pitting Corrosion of Metals*, National Association of Corrosion Engineers, Houston, TX, p. 39.
5   Szklarska-Smialowska, Z., 1986, *Pitting Corrosion of Metals*, National Association of Corrosion Engineers, Houston, TX.
6   NACE, 1982, *The Forms of Corrosion: Recognition and Prevention*, Handbook 1, Houston, TX.
7   NACE, 1984, *Corrosion Basics, An Introduction*, National Association of Corrosion Engineers, Houston, TX, p. 27.
8   NACE, 1984, *Corrosion Basics, An Introduction*, National Association of Corrosion Engineers, Houston, TX, p. 31.

9   ASM Metals Handbook, 1987, Vol. 13, *Corrosion*, ASM International, Metals Park, OH, p. 29.
10 Piron, D.L., 1991, *The Electrochemistry of Corrosion*, National Association of Corrosion Engineers, p. 77.
11 ASM Metals Handbook, 1987, Vol. 13, *Corrosion*, ASM International, Metals Park, OH, p. 104.
12 ASM Metals Handbook, 1987, Vol. 13, *Corrosion*, ASM International, Metals Park, OH, p. 111.
13 NACE, 1984, *Corrosion Basics, An Introduction*, National Association of Corrosion Engineers, Houston, TX, p. 97.
14 ASM Metals Handbook, 1987, Vol. 13, *Corrosion*, ASM International, Metals Park, OH, p. 109.
15 NACE, 1984, *Corrosion Basics, An Introduction*, National Association of Corrosion Engineers, Houston, TX, p. 34.
16 NACE, 1984, *Corrosion Basics, An Introduction*, National Association of Corrosion Engineers, Houston, TX, p. 35.
17 NACE, 1984, *Corrosion Basics, An Introduction*, National Association of Corrosion Engineers, Houston, TX, p. 111.
18 NACE, 1984, *Corrosion Basics, An Introduction*, National Association of Corrosion Engineers, Houston, TX, p. 113.
19 NACE, 1984, *Corrosion Basics, An Introduction*, National Association of Corrosion Engineers, Houston, TX, p. 118.
20 Dillon, C.P., 1986, *Corrosion Control in the Chemical Process Industries*, McGraw-Hill, New York, p. 170.
21 Dillon, C.P., 1986, *Corrosion Control in the Chemical Process Industries*, McGraw-Hill, New York, p. 174.
22 NACE, 1984, *Corrosion Basics, An Introduction*, National Association of Corrosion Engineers, Houston, TX, p. 102.
23 NACE, 1984, *Corrosion Basics, An Introduction*, National Association of Corrosion Engineers, Houston, TX, p. 276.
24 ASM Metals Handbook, 1987, Vol. 13, *Corrosion*, ASM International, Metals Park, OH, p. 37.
25 Lorbeer, P., and Lorenz, K., 1980, 'The kinetics of iron dissolution and passivation in solutions containing oxygen', *Electrochim. Acta.* **25**(4) 375.
26 Uhlig, H.H., and Revie, R.W., 1985, *Corrosion and Corrosion Control*, John Wiley & Sons, New York, p. 21.
27 NACE, 1984, *Corrosion Basics, An Introduction*, National Association of Corrosion Engineers, Houston, TX, p. 37.
28 NACE, 1984, *Corrosion Basics, An Introduction*, National Association of Corrosion Engineers, Houston, TX, p. 38.
29 Szklarska-Smialowska, Z., 1986, *Pitting Corrosion of Metals*, National Association of Corrosion Engineers, Houston, TX, p. 301.
30 Uhlig, H.H., 1942, *Trans. American Society Metals* **30**(12) 949.
31 Fontana, M., 1986, *Corrosion Engineering*, McGraw-Hill, New York.
32 Uhlig, H.H., 1948, *Corrosion Handbook*, John Wiley & Sons, New York, p. 481.
33 Peckner, D., and Bernstein, I.M., 1977, *The Handbook of Stainless Steels*, McGraw-Hill, New York, p. 16-2.

34 Okomoto, G., 1973, 'Passive film of 18–8 stainless steel structure and its function', *Corr. Sci.* **13** 471–9.

35 Licina, G., 1988, *Sourcebook for Microbiologically Influenced Corrosion in Nuclear Power Plants*, EPRI-NP-5580, Electric Power Research Institute, Palo Alto, CA.

36 Poyet, P., and Desestret, A., 1974, *Mem. Sci. Rev. Met.* **121** 467.

37 Lorenz, K., and Medawar, G., 1969, *Tyssenforschung* **1** 97.

38 Forchhammer, P., and Engell, H.J., 1969, *Werkst. Korros.* **20** 1.

39 Wilde, B.E., and Armijo, J.S., 1967, *Corrosion* **23**(7) 208.

40 Szklarska-Smialowska, Z., 1986, *Pitting Corrosion of Metals*, National Association of Corrosion Engineers, Houston, TX, p. 78.

41 Steinemann, S., 1968, *Mem. Sci. Rev. Met.* **65** 615.

42 Devine, T.M., 1979, *J. Electrochem. Soc.* **38**(2) 63.

43 Hronsky, P., and Duquette, D., 1982, *Corrosion* **38**(2) 63.

44 Golovina, G.V., Florianovich, G.M., and Kolotrykin, Y., 1965, *Elektrokhima* **1** 12.

45 NACE, 1984, *Corrosion Basics, An Introduction*, National Association of Corrosion Engineers, Houston, TX.

46 Betz, 1991, *Handbook of Industrial Water Conditioning*, Betz laboratories, Trevose, PA, p. 175.

47 NACE, 1984, *Corrosion Basics, An Introduction*, National Association of Corrosion Engineers, Houston, TX, p. 348.

48 NACE, 1984, *Corrosion Basics, An Introduction*, National Association of Corrosion Engineers, Houston, TX, p. 12.

49 ASM Metals Handbook, 1987, Vol. 13, *Corrosion*, ASM International, Metals Park, OH, p. 39.

50 Szklarska-Smialowska, Z., 1986, *Pitting Corrosion of Metals*, National Association of Corrosion Engineers, Houston, TX, p. 147.

51 Streicher, M.A., 1974, *Corrosion* **30**(2) 77.

52 Brigham, R.J., and Tozer, E.W., 1974, *J. Electrochem. Soc.* **121** 1192.

53 Szklarska-Smialowska, Z., 1986, *Pitting Corrosion of Metals*, National Association of Corrosion Engineers, Houston, TX, p. 157.

54 Fontana, M., 1986, *Corrosion Engineering*, McGraw-Hill, New York, p. 71.

55 Szklarska-Smialowska, Z., 1986, *Pitting Corrosion of Metals*, National Association of Corrosion Engineers, Houston, TX, p. 189.

56 Szklarska-Smialowska, Z., Szummer, A., and Janik-Czachor, M., 1970, *Brit. Corros. J.* **15**(5) 159.

57 Poyet, P., and Desestret, A., 1975, *Mem. Sci. Rev. Met.* **72** 133.

58 Garner, A., 1979, *Corrosion* **35**(3) 108.

59 Szklarska-Smialowska, Z., 1986, *Pitting Corrosion of Metals*, National Association of Corrosion Engineers, Houston, TX, p. 194.

60 Stalder, F., and Duquette, D., 1975, Proc. 6th Int. Cong. Met. Corros., Sydney, Australian Corrosion Association, Parkville, Australia, paper 1–33.

61 Buchanan, R.A., and Stansbury, E.E., 1991, 'Fundamentals of coupled electrochemical reactions as related to microbially influenced corrosion', in *Microbially Influenced Corrosion and Biodeterioration*, Dowling, N.J., *et al.*, ed., National Association of Corrosion Engineers, Houston, TX, 1–11.

62 Characklis, W.G., Little, B.J., and McCauley, M.S., 1989, 'Biofilms and

their effect on local chemistry', *Microbial Corrosion: 1988 Workshop Proceedings*, EPRI ER-6345, Energy Power Research Institute, Palo Alto, CA.

63 Stoecker, J.G., and Pope, D.H., 1986, 'Study of biological corrosion in high temperature demineralized water', *Materials Performance* **25**(6) 58.

64 Jones, J.G., 1986, 'Iron transformations by freshwater bacteria', in *Advances in Microbial Ecology*, Marshall, K.C., ed., Vol. 9, p. 149, Plenum, NY.

65 ASM Metals Handbook, 1987, Vol. 13, *Corrosion*, ASM International, Metals Park, OH, p. 118.

66 Borenstein, S.W., 1988, *Materials Performance* **27**(8) 62.

67 Uhlig, H.H., and Revie, R.W., 1985, *Corrosion and Corrosion Control*, John Wiley & Sons, New York, p. 171.

68 Vernon, W., 1927, *Trans. Faraday Soc.* **23** 113.

69 *Chemical Engineering News*, January 4, 1971, p. 29.

70 NACE, 1984, *Corrosion Basics, An Introduction*, National Association of Corrosion Engineers, Houston, TX, p. 222.

71 ASM Metals Handbook, 1987, Vol. 13, *Corrosion*, ASM International, Metals Park, OH, p. 513.

72 ASM Metals Handbook, 1987, Vol. 13, *Corrosion*, ASM International, Metals Park, OH, p. 512.

73 Uhlig, H.H., and Revie, R.W., 1985, *Corrosion and Corrosion Control*, John Wiley & Sons, New York, p. 176.

74 Bennett, L.H., 1978, National Bureau of Standards, Spec. Publ. 511-1, Washington, DC, p. 16.

75 NACE, 1984, *Corrosion Basics, An Introduction*, National Association of Corrosion Engineers, Houston, TX, p. 19.

76 Uhlig, H.H., and Revie, R.W., 1985, *Corrosion and Corrosion Control*, John Wiley & Sons, New York, p. 178.

77 Tomashov, N.D., and Mikailovsky, Y.N., 1959, *Corrosion* **15** 77t.

Chapter 5

Case histories

Many papers, articles and books have discussed MIC failures.[1-5] The illustrated case histories from NACE, *A Practical Manual on Microbiologically Influenced Corrosion*, are recommended.

While case histories can help us visualize typical scenarios, they cannot necessarily increase our fundamental understanding of the failure mechanisms, given their inherent complexity. Consider the existence of mixed-mode failures, in which one mechanism or mode of failure is initially dominant, superceded by the second, later in the failure. Information gathered after a failure is often incomplete, leading to disagreement among investigators about the cause-and-effect relationship of corrosion, particularly regarding MIC. In addition, some system failures are 'corrected', only to recur later.

Ideally, we can derive considerable information from case histories about what went wrong, and why. Sometimes what we really want to know is what went right and why, information that may be unavailable. Why did one system fail, and the one beside it that operated 'the same' have no problems? Why did one weld pit and leak and the ones on either side have no corrosion? Often, our questions can be answered only by speculation.

With those caveats noted, this chapter presents case histories of MIC failures, taking into account symptoms, methods of investigation, possible corrosion mechanisms and corrective actions.

5.1 Carbon steels

For the carbon and low alloy steels, MIC-related problems usually consist of tubercles with pitting underneath. In addition to problems with pitting, general corrosion or reduced flow, hydraulic-type problems and problems of plugging are very common with carbon steel.

Carbon steels are frequently used for raw, untreated water, which means that metals are exposed to an extremely wide variety of aerobic and anaerobic microorganisms. As discussed previously, this environment results in a complex set of variables that contributes to corrosion. The association of microorganisms with steel corrosion has been discussed for many years.[1,2]

5.1.1 Theories

Microbiological influences on corrosion are in most cases associated with colonies of organisms, scale, debris and exopolymers forming local deposits on a metal surface. One popular theory is that the deposits form and often initiate pits on the metal surface. These pits then propagate and corrode the metal surface.

An alternative theory is that the metal surface is not homogeneous and that pits initiate, resulting in patchy regions where the corrosion resistance of the surface is lower than the surrounding surface. Corrosion begins to occur on these regions. The electrochemical nature of corrosion may induce electron transfers; the electron donors and the electron acceptors attract microbes to the site. The microbes attach and colonize, creating a more complicated environment and a 'different type' of corrosion cell.

5.1.2 Pitting under tubercles

Pitting is usually associated with tubercles, occasionally in isolated areas. Pitting results from concentration cells set up by tubercles. The influences of chemical reactions are also a factor in forming tubercles. Influences such as oxygen depletion, concentration of chlorides and reduction of sulphates increase the susceptibility to corrosion in specific materials and environments, and combined with microbial activity, often result in MIC. The formation of tubercles seems to increase the probability of pitting's occurrence. Mechanical aspects such as crevices (gaskets and backing rings) or fabrication (poor quality welds) produce conditions for micro-organism growth and localized corrosion.[5] All these situations (Fig. 5.1) produce environments conducive to microbial growth and accelerated corrosion due to MIC.[6,7]

Aerobic bacteria such as *Gallionella* and *Leptothrix* are generally thought to contribute to corrosion by forming differential aeration cells. As oxygen-starved conditions develop, corrosion occurs underneath. In addition, the microorganisms formed on the surface of the metal seem to concentrate chlorides and manganese ions in their metabolic process.

These organisms (*Gallionella* and *Leptothrix*) are suspected of converting the soluble ferrous ion to the less soluble ferric iron, a chemical reaction that results in ferric hydroxide (Fig. 5.2).[5,8] The result of these reactions is thought to encourage the growth of the microorganisms and provide an environment for other organisms. When microorganisms nest together, they form a consortium, as described earlier. Aerobic organisms form and grow on the water side of the tubercle, and anaerobic organisms

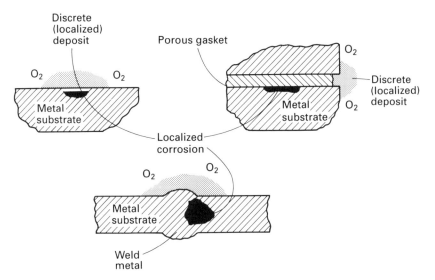

5.1 The three most common forms of microbial corrosion (after Tatnell, R.E., 1985, 'Experimental methods in biocorrosion', in *Biologically Influenced Corrosion*, Dexter, S.C., ed., NACE, Houston, TX, reproduced with permission).

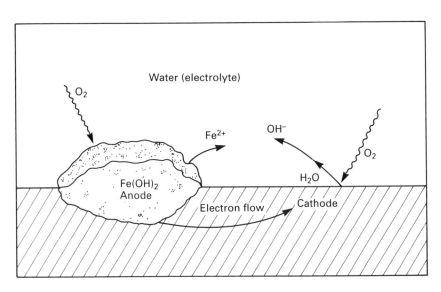

5.2 Corrosion cell.

do so on the metal surface. The tubercle protects the colony of organisms. In addition, the tubercle resists the effectiveness of the corrosion inhibitors to biocides.

Anaerobic bacteria that influence corrosion, such as the sulphate-reducing bacteria, have been studied for years.[3–5, 9–16] The exact mechanism by which the corrosion occurs is still being debated. As Licina reminds us, SRB activity can be significant in aerated environments while sheltered from the bulk environment by tubercles. Extremely rapid attack, at rates of 100 mils per year or more, can result under tubercles, as a result of the SRBs.[5, 17, 18]

5.1.3   Reduced flow and hydraulic effects

In addition to pitting, reduced flow and hydraulic problems can occur. Clogging results when thick deposits of corrosion block the fluid flow.[4, 5, 13, 17, 19] Deposits with very low densities (10%) have been reported in raw water applications.[17]

These flow-related effects can cause problems. Pressure drop increases dramatically, and, in some cases, cooling flow to critical equipment will be insufficient.[19] Removing the deposit removes the flow difficulties. But, unless the cause of the corrosion problem has been removed too, deposits will begin to reform immediately.

In a further problem, accelerated attack of the pipe wall will occur in areas previously covered by the massive deposits, which can affect the integrity of the pipe, and potentially damage equipment function (for example, water flowing to lube oil coolers). The water will not flow properly because of the flow restrictions and clogging; the lube oil no longer cools properly, the oil gets too hot, and the unit shuts down.

5.2   Copper-based alloys

Copper-based alloys are used in equipment such as heat exchangers, pumps, valves and condensers. The copper-based alloys commonly used are 90–10 and 70–30 copper–nickel, brasses and aluminium bronzes, as well as the admiralty brasses (Fig. 5.3). Little *et al.* have documented localized corrosion of copper alloys by SRBs in marine environments.[20–23]

All these alloys are susceptible to MIC, a prevalent notion that copper-based alloys are immune to MIC not withstanding. Old literature often states that copper is toxic to common fouling organisms. Regrettably, some technical literature has unwittingly blended fact and fiction. Geesey *et al.* demonstrated that extracellular polymers play a role in the corrosion of copper alloys.[24–26] The role of the extracellular polymers is discussed more fully in Chapter 2. Differential aeration, selective leaching, under-

5.3   SEM of bacteria and exoplasm brass.

deposit corrosion and cathodic depolarization have all been reported as mechanisms for MIC of copper alloys.

Copper alloys are usually considered very resistant to fouling.[27, 28] However, as Licina has discussed, pitting, plug dealloying and ammonia-induced stress corrosion cracking have been reported.[5, 24, 29, 30] The one exception to the copper alloys' nonsusceptibility to MIC seems to be the arsenical coppers, which are no longer commonly used because of environmental concerns. It is thought that the adverse effects on copper alloys result from the metabolic products of particular organisms attacking the corrosion film.

In terms of corrosion, copper-based alloys are resistant to neutral and slightly alkaline solutions, the exception being ammonia, which can cause stress corrosion cracking and may sometimes cause general attack. In terms of microorganisms, ammonia is a common organism waste product, so when it attacks copper-based alloys, whether the ammonia comes from organisms or other sources is immaterial.

Organic acids attack copper and are produced by many organisms – a key feature in the problems faced when distinguishing between the different aspects of MIC. Sulphides are also particularly corrosive to many

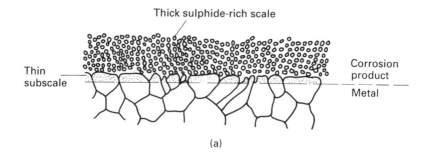

5.4 Schematics of (a) thick sulphide rich scale on copper alloy and (b) disruption of sulphide film by introduction of aerated sea water (after Syrett, B.C., 1980, *Corrosion/80*, paper no. 30, NACE, Houston, TX, reproduced with permission).

metals, so sulphides produced by sulphate-reducing bacteria are considered particularly potent.[29,31] As Licina has noted, sulphate-reducing bacteria are thought to produce hollow, hemispherical tubercles. The pits underneath are of essentially the same size and shape.[32–34] A significant increase in pitting of copper was produced with sulphide contents of less than 0.2 ppm.[29,32]

Licina discusses how an oxide is produced upon exposure to sulphide-containing solutions, which commonly occurs during wet lay-up.[32] Accelerated corrosion occurs after dissolved oxygen in the water touches this film. Little *et al.* discuss the sulphide film on copper alloys in seawater.[20] These films are fragile and composed of a complex system of copper sulphides.[35] Figure 5.4 illustrates the scale attachment and spalling.[36]

Erosion–corrosion can plague copper alloys and may sometimes be associated with MIC. High local velocity can repeatedly strip the protective film on the surface of copper alloys, and the metal rapidly degrades.

An example is often seen in heat exchanger tubes: corrosion or debris blocks the tube, the velocity increases, and the tube deteriorates.

Pope *et al.* proposed that the following microbial products accelerate localized attack: CO_2, H_2S, NH_3, organic and inorganic acids; metabolites that act as depolarizers; and sulphur compounds such as mercaptans, sulphides and disulphides.

5.3 Nickel-based alloys

Nickel-based alloys such as Monels™, Incoloys™ and Inconels™ are generally specified for applications that require a high degree of corrosion resistance. They are generally thought to be less susceptible to MIC, although a new MIC case history discusses Monel 400.[37]

Licina discusses that, in relation to nuclear power plants, most light water reactors (LWR) applications of the nickel-based alloys materials are restricted to critical areas such as pressurized water reactors (PWR) and steam generators.[5] He writes that even with the high resistance to pitting in oxidizing environments, including high chloride environments, that is usually exhibited by these materials, several case histories imply that their resistance to MIC may not be significantly greater than that of stainless steel.[33, 34, 38]

Tatnall discusses a case of pitting in nickel tubes and shallow pitting in Alloy 400 (UNS N04400) and Alloy B (UNS N10001).[3] He concluded that these failures may be related to MIC. SRBs were found in the cooling water.

Pope *et al.* discuss bacteria oxidizing Cr^{+3} to Cr^{+6}.[39, 40] The Cr^{+6} form is more soluble, but thought to be more toxic to organisms. Some bacteria use CrO_3 as a final acceptor for electrons generated during metabolism, resulting in a reduction of the Cr^{+3} to Cr^{+6} form, which may alter the passive film of chromium oxide on chromium-containing alloys, resulting in corrosion.[39, 41] Diekert and Ritter found that some bacteria, *Acetobacterium woodii*, use nickel in a reaction in which H_2 plus CO_2 is converted into acetic acid.[42]

It is interesting to note that after a MIC failure in a chemical process plant, 90–10 copper nickel was specified as the replacement for nickel.[3]

5.4 Stainless steels

The 18% chromium 8% nickel composition of austenitic stainless steels, commonly referred to as 18–8 stainless (such as Type 304), are often used in constructing piping systems. For many applications, such as nuclear power plants, they are used for their high leak-before-break properties (meaning they have high toughness).[5] Stainless steels are also commonly used in high purity water environments, such as reactor coolant systems, emergency systems, reactor auxiliary systems, feedwater and condensers.

Stainless steel often seems vulnerable to MIC problems during startup. This phase of preoperational testing may use untreated or potable water. For some failures, the water was potable well water but contained many naturally occurring microorganisms. These organisms, such as the iron-oxidizing and iron-depositing bacteria, will not make people sick, but seem to be frequently associated with MIC failures.

Stoecker and Pope found that deionized water can also contain micro-organisms.[43] Their case history discussed a single microbiological species at elevated temperature in relatively pure deionized water, resulting in MIC producing pitting failures in 304 stainless steel.

5.4.1   Pitting at weldments

As discussed earlier, MIC of the austenitic stainless steels is often characterized by pitting, usually at or adjacent to weldments. Such pitting results in throughwall pitting with small leaks, and is a common consequence of MIC, particularly to austenitic stainless steel weldments.

The US Nuclear Regulatory Commission has expressed concern over corrosion of austenitic stainless steel used in piping systems in nuclear power plants. US Nuclear Regulatory Commission I&E bulletin 85-30, INPO Significant Event Report 73-84 and the US Nuclear Regulatory Commission Generic Letter 89-13 address the specific concerns of MIC and failures in nuclear power plants.

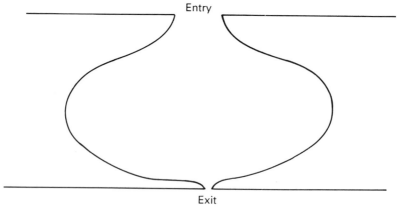

5.5 'Tunnelling' pit observed in stainless steel (after Licina, G.J., 1988, *Sourcebook for Microbiologically Influenced Corrosion in Nuclear Power Plants*, EPRI NP-5580, Electric Power Research Institute, Palo Alto, CA, reproduced with permission).

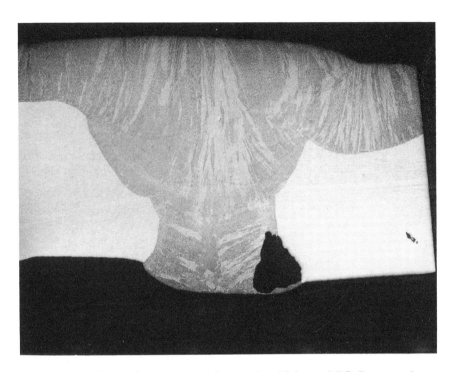

5.6 Pitting of austenitic stainless steel weld due to MIC. Base metal 316L, metal ER308. 5×.

Pits often exhibit small deep holes, as illustrated in Fig. 5.5.[3,5,33,38,39,44] Tatnall cautions that the same pit shape can be produced when manganese-rich deposits contact chlorinated water, whether bacteria are present or not, which appears to result from a reaction involving chloride and permanganate ions.

Pits may also be associated with attack at the weld (Fig. 5.6, 5.7), generally near the fusion line, or with sensitization in the heat-affected zone. Base metal attack is much less common.[33,45-47] The two-phase weld metal appears to be the most susceptible area, although the relative susceptibility of the austenite and delta ferrite phases has not been clearly defined. The interfacial region between the two phases may also be susceptible. As shown in the referenced papers, in some situations the austenite is preferentially attacked.[3,48-51] In other situations, the delta ferrite is removed and the austenite is unattacked.[3,51,52]

According to Stein, nickel–chrome–molybdenum weld metal *may* exhibit greater resistance to stainless steel weld metal.[53] Stein observed attack in HAZs of a Type 316 stainless steel component, while the adjacent Inconel 625 weld metal was unattacked.

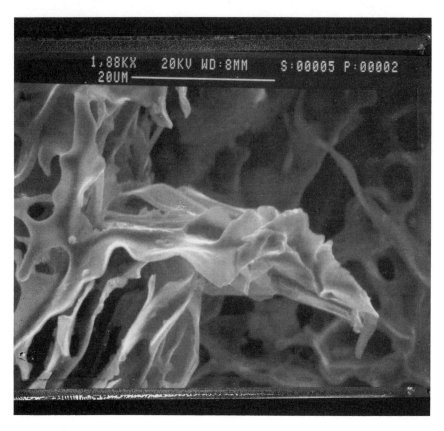

5.7 SEM photomicrograph showing matrix remaining after preferential corrosion of the S ferrite in a type 316 stainless steel weldment.

As discussed in Section 2.3, the iron-oxidizing bacteria, such as *Gallionella* and *Siderocapsa*, as well as the iron-depositing bacteria, are commonly associated with MIC of austenitic stainless steels. They have been associated with MIC with the pit morphology described above (selective phase attack). The investigator should be cautious with theories about the iron-oxidizing and iron-depositing bacteria. There is no direct correlation between the presence of the bacteria and the deterioration due to corrosion at this time. It may simply be that the bacteria are attracted to the metal surface because of the metal ions. The metal ions are in solution at the surface due to some other initiating corrosion mechanism, and the bacteria are attracted to this site.[54]

SRBs' activity has sometimes been associated with stainless steels. Pitting is the primary form of corrosion and the pit morphology is different in stainless steels. In general, the pits were round-bottomed shallow pits rather than bottle-shaped. The investigator is cautioned about giving too much weight in diagnosis to pit morphology. It may be an indicator, but not the only one.

5.4.2   Super stainless steels

The family of stainless steels known as 'super' stainless steel include the duplex stainless steels and 6% molybdenum stainless steel alloys, such as AL-6XN™ (UNS N08367) (trademark of Allegheny Ludlum Steel, Brackenridge, PA). These 6% molybdenum alloys are more resistant to pitting than the 300 series austenitic stainless steels. They have some relatively minor problems with crevice corrosion and appear to behave similarly to the 300 series austenitic stainless steels in the effect of heat tint on slightly reducing corrosion resistance.[25, 47]

Ferralium 255 (UNS S32550) is a duplex stainless steel with 25.5% chromium and 5.5% nickel, with about 3% molybdenum and other alloying elements. Licina considers it in relation to service in situations where MIC has occurred, noting its resistance to pitting in acidic and chloride environments.[55] Stoecker noted corrosion of Ferralium 255 and an association with MIC in stagnant seawater.[38]

Scott and Davies described corrosion in a heat exchanger used for cooling sulphuric acid in brackish seawater applications.[56] The failure to the 904L tubes appeared to result from MIC associated with SRBs. The 904L (UNS N08904) has a nominal composition of 20% chromium, 25% nickel, 4.5% molybdenum, 1.5% copper, 0.02% maximum carbon and the balance iron, and is thought to be resistant to most corrosion problems associated with seawater.

5.5   Other materials

Other common materials of construction include titanium, aluminium and concrete.

5.5.1   Titanium

Titanium is susceptible to biofouling (undesirable biological growth on a surface).[27, 55, 57, 58] Glass surfaces can biofoul.

Pope states that, in theory, sulphuric acid is produced by *Thiobacillus* and that corrosion may occur underneath on titanium surfaces.[39] However, there are no documented case histories of MIC of titanium and titanium alloys.[25] Schultz reviewed mechanisms for MIC and titanium's corrosion behaviour under various conditions.[59] He concluded that at temperatures below 100 °C (212 °F), titanium is invulnerable to sulphur-oxidizing bacteria (SOB), SRBs, acid-producing bacteria, differential aeration cells, chloride concentration cells and hydrogen embrittlement. In laboratory studies, Little *et al.* did not observe MIC in the presence of SRBs or iron/sulphur-oxidizing bacteria at either mesophilic (23 °C; 73 °F) or thermophilic (70 °C; 158 °F) temperatures.[58]

5.5.2   Aluminium

Aluminium's corrosion resistance is due to a thin adherent film of aluminium oxide on the metal surface. This film is naturally occurring, formed in air, and is approximately 2–10 nm thick. Thicker films may be made by anodizing, by producing a thick (0.005–0.03 mm; 0.2–1.2 mils) oxide coating of aluminium oxide on the surface of the alloy by an electrolytic process.[60] The protective film is susceptible to corrosion by attack from halide ions such as chlorides. This susceptibility to localized corrosion appears to make aluminium alloys vulnerable to MIC.

Most reports are for aluminium (99%), 2024 and 7075 alloys used in aircraft or in underground fuel storage tanks.[20, 25, 61–63] Often, the localized corrosion is attributed to MIC in the water phase of fuel–water mixtures in tank bottoms or at the fuel–water or fuel–air interfaces. Contaminants in the fuel such as surfactants, water and water-soluble salts may contribute to bacterial growth. Little *et al.* note that two mechanisms for MIC of aluminium alloys have been documented: the production of water–soluble acids by bacteria and fungi, and the formation of differential aeration cells.[20]

Schmitt reported severe MIC of a finned aluminium alloy heat exchanger after several months' exposure to potable water.[64] Metallographic examination of corrosion sites showed cubic pitting, indicative of an acid-chloride attack. Additional problems due to MIC were reported for aluminium screens in a solar energy system covered with a biofilm.[64]

Aluminium–magnesium alloys (5000 series) used for marine applications resist uniform corrosion; however, some are susceptible to pitting, intergranular attack, exfoliation and stress corrosion cracking. Crooker observed that biofilm formation in natural seawater may play a role in the basic mechanism of crack formation.[65]

5.5.3   Concrete

Corrosion of concrete can occur. Concrete is used extensively underground, often with reinforcing bars.

Soil chemistry as well as groundwater must be considered for underground applications. *Corrosion Basics, An Introduction* discusses corrosion in soils and addresses concrete.[66] According to their guidelines, the soil must be judged aggressive if the following are true:

- The pH is less than 6.
- The pH is neutral, but the hydrogen ion is available by exchange.

- The sulphate or sulphide content is high.
- The magnesia content is high.

Dry soil or well-drained sandy conditions are usually considered ideal and result in little degradation.

Concrete is susceptible to attack by acids. Any conditions that produce acid should be considered detrimental to the concrete and capable of accelerating corrosion. Sand *et al.* found that sulphur-oxidizing bacteria degrade concrete.[67] Van Mechelen and Polder found MIC to be a serious problem in concrete sewer pipes.[68] They determined the root cause of failure to be sulphuric acid attack to the concrete. The mechanism was

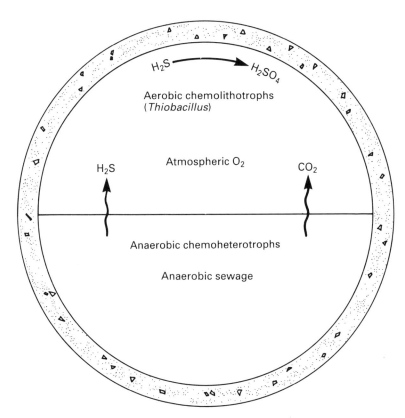

5.8 Crown corrosion in a concrete sewer. Anaerobic chemoheterotrophs produce H_2S, which is absorbed in condensate on the walls above the water line. Aerobic chemolithotrophs oxidize the reduced sulphur to SO_4^{2-}, forming H_2SO_4, which attacks and erodes the concrete, often to the point of structural failure (figure is from *Elements of Bioenvironmental Engineering* by Gaudy and Gaudy, and is used by permission of Engineering Press, Inc. the copyright owner).

bacteria reducing sulphate to sulphide and other sulphur compounds; the sulphide was then released to the sewage and, when the sewage contained sufficient oxygen, it oxidized and formed sulphuric acid. In other cases, the sulphide remains in reduced form and is released as hydrogen sulphide which is oxidized by meeting a moist environment and the bacteria, such as *Thiobacilli*, and then forms sulphuric acid (Fig. 5.8).[69]

Hall found MIC in concrete in wastewater collection and treatment systems and describes exposures of various concretes to evaluate their resistance to MIC.[70] Brock and Sand found MIC in concrete from the production of sulphuric acid, as well as in natural sandstone in historic buildings from the production of nitric acid.[71]

Suggested reading

1 Miller, J.D.A., and Tiller, A.K., 1970, *Microbial Aspects of Metallurgy*, American Elsevier, New York.
2 NACE, 1990, *TPC 3 Microbiologically Influenced Corrosion and Bio-fouling in Oilfield Equipment*, National Association of Corrosion Engineers, Houston, TX.
3 Dexter, S.C., ed., 1986, *Biologically Influenced Corrosion*, NACE Reference Book No. 8, National Association of Corrosion Engineers, Houston, TX.
4 Licina, G.J., 1989, *Microbial Corrosion: 1988 Workshop Proceedings*, EPRI ER-6345, Electric Power Research Institute, Palo Alto, CA.
5 Pope, D.H., 1986, *A Study of Microbiologically Influenced Corrosion in Nuclear Power Plants and a Practical Guide for Countermeasures*, EPRI NP-4582, Electric Power Research Institute, Palo Alto, CA.
6 Licina, G.J., 1988, *Sourcebook for Microbiologically Influenced Corrosion in Nuclear Power Plants*, EPRI NP-5580, Electric Power Research Institute, Palo Alto, CA.
7 Pope, D.H., Duquette, D., Wagner, P.C., and Johannes, A.H., 1989, *Microbiologically Influenced Corrosion: A State-of-the-Art Review*, MTI *Publication No. 13*, Materials Technology Institute of the Chemical Process Industries, National Association of Corrosion Engineers, Houston, TX.
8 Licina, G.J., 1988, *Detection and Control of Microbiologically Influenced Corrosion*, EPRI NP-6815-D, Electric Power Research Institute, Palo Alto, CA.
9 Mittleman, M., and Geesey, G., 1987, *Biological Fouling of Industrial Water Systems: A Problem Solving Approach*, Water Micro Associates, San Diego, CA.
10 Dowling, N.J., Mittleman, M., and Danko, J., ed., 1990, *Microbially Influenced Corrosion and Biodeterioration*, National Association of Corrosion Engineers, Houston, TX.

References

1 Miller, J.D.A., and Tiller, A.K., 1970, *Microbial Aspects of Metallurgy*, American Elsevier, New York.

2   Iverson, W.P., 1974, *Microbial Iron Metabolism*, Academic Press, New York.

3   Tatnall, R.E., 1981, 'Fundamentals of bacteria induced corrosion', *Materials Performance* **20**(9) 32.

4   Pope, D.H., 1986, 'A study of microbiologically influenced corrosion in nuclear power plants and a practical guide for countermeasures', EPRI NP-4582, Electric Power Research Institute, Palo Alto, CA.

5   Licina, G.J., 1988, *Sourcebook for Microbiologically Influenced Corrosion in Nuclear Power Plants*, EPRI NP-5580, Electric Power Research Institute, Palo Alto, CA, p. 4-4.

6   ASM Metals Handbook, 1987, Vol. 13, *Corrosion*, ASM International, Metals Park, OH, p. 115.

7   Tatnall, R.E., 1986, 'Experimental methods in biocorrosion' in *Biologically Influenced Corrosion*, Dexter, S.C., ed., National Association of Corrosion Engineers, Houston, TX.

8   Tiller, A.K., 1983, 'Is stainless steel susceptible to microbial corrosion?' in Proceedings of the conference sponsored and organized by the National Physics Laboratory and The Metals Society, NPL, Teddington.

9   Kobrin, G., 1976, 'Corrosion by microbiological organisms in natural waters', *Materials Performance* **15**(6) 38.

10 Crombie, D.J., Moody, G.J., and Thomas, J.D.R., 1980, 'Corrosion of iron by sulfate-reducing bacteria', *Chemistry and Industry* **21** 500.

11 Saunders, P.F., and Hamilton, W.A., 1986, 'Biological and corrosion activities of sulfate-reducing bacteria in industrial plants', in *Biologically Induced Corrosion*, Dexter, S.C., ed., National Association of Corrosion Engineers, Houston, TX.

12 Bibb, M., and Hartman, K.W., 1984, 'Bacterial corrosion', *Corrosion and Coatings South Africa*, October, 12.

13 McDougal, J., 1966, 'Microbial corrosion of metals', *Anti-Corrosion*, August, 9.

14 Tiller, A.K., 1986, 'A review of the European research effort on microbial corrosion between 1950 and 1984', in *Biologically Induced Corrosion*, Dexter, S.C., ed., National Association of Corrosion Engineers, Houston, TX.

15 Hamilton, W.A., and Maxwell, S., 1986, 'Biological and corrosion activities of sulfate-reducing bacteria', in *Biologically Induced Corrosion*, Dexter, S.C., ed., National Association of Corrosion Engineers, Houston, TX.

16 Williams, R.E., Ziomek, E., and Martin, W.G., 1986, 'Surface stimulated increases in hydrogenase production by sulfate reducing bacteria', in *Biologically Induced Corrosion*, Dexter, S.C., ed., National Association of Corrosion Engineers, Houston, TX.

17 Metell, H.M., 1985, 'ASME pressure vessel piping', in Proceedings of the 1985 Pressure Vessel Piping Conference, New Orleans, LA, June.

18 Smith, J.P., and Surinach, P.P., 1987, 'Biocide selection and optimization in IIAPCD's Seawater Injection Project', *Corrosion/87*, paper no. 366, National Association of Corrosion Engineers, Houston, TX.

19 Bowman, C.F., and Bain, W.S., 1980, *Power Engineering*, August, 73.

20 Little, B.J., Wagner, P., and Mansfeld, F., 1991, 'Microbiologically influenced corrosion of metals and alloys', *International Materials Review* **36**(6) 253.

21 Little, B., Wagner, P., Ray, R., and McNeil, M., 1990, *Mar. Technol. Soc. J.* **24** 10.

22 Little, B.J., Wagner, P., and Jacobus, J., 1988, *Materials Performance* **27**(8) 56.

23 Little, B.J., Wagner, P., Jacobus, J., and Janus, L., 1989, *Estuaries* **12**(3) 138.

24 Pope, D.H., Duquette, D., Wagner, P.C., and Johannes, A.H., 1984, *Materials Performance* **23**(4) 14.

25 Wagner, P., Little, B.J., and Borenstein, S.W., 1992, 'The influence of metallurgy on microbiologically influenced corrosion', *Proc. Marine Technology Society*, Washington, DC, p. 1044.

26 Geesey, G.G., Mittleman, M.W., Iwaoka, T., and Griffiths, P.R., 1986, *Materials Performance* **25**(6) 37.

27 Lewis, R.O., 1982, 'The influence of biofouling countermeasures on corrosion of heat exchanger materials in seawater', *Materials Performance*, **21**(9) 31.

28 Videla, H., and DeMele, M.F.L., 1987, 'Microfouling of several metal surfaces in polluted seawater and its relation with corrosion', *Corrosion/87*, paper no. 365, National Association of Corrosion Engineers, Houston, TX.

29 EPRI CS-4562, 1986, 'Effects of sulfide, sand, temperature and cathodic protection on corrosion of condensers' Electric Power Research Institute, Palo Alto, CA.

30 Alanis, I., Berardo, L., De Cristofaro, N., Moina, C., and Valentini, C., 1986, 'A case of localized corrosion in underground brass pipes', in *Biologically Influenced Corrosion*, Dexter, S.C., ed., National Association of Corrosion Engineers, Houston, TX.

31 Staffeldt, E.E., and Calderon, O.T., 1969, *Div. Ind. Microbiol.* **8** 321.

32 Licina, G.J., 1988, *Sourcebook for Microbiologically Influenced Corrosion in Nuclear Power Plants*, EPRI NP-5580, Electric Power Research Institute, Palo Alto, CA.

33 Kobrin, G., 1976, *Materials Performance* **15**(7) 38.

34 Tatnall, R.E., 1981, *Materials Performance* **20**(8) 41.

35 Ribbe, P.H., 1976, 'Sulfide mineralogy', CS-58-CS-76, Mineralogical Society of America, Washington, DC.

36 Syrett, B.C., 1980, *Corrosion/80*, paper no. 30, National Association of Corrosion Engineers, Houston, TX.

37 Gouda, V.K., Banat, I.M., Riad, W.T., and Mansour, S., 1993, *Corrosion* **49**(1) 63.

38 Stoecker, J.G., 1984, 'Guide for the investigation of microbiologically influenced corrosion', *Materials Performance* **23**(8) 48.

39 Pope, D.H., Duquette, D., Wagner, P.C., and Johannes, A.H., 1989, *Microbiologically Influenced Corrosion: A State-of-the-Art Review*, MTI Publication No. 13, Materials Technology Institute of the Chemical Process Industries, National Association of Corrosion Engineers, Houston, TX.

40 Bartlett, R., 1979, *J. Envir. Quality* **8** 31.

41 Bopp, L., 1980, PhD Thesis, Rensselaer Polytechnic Institute, Troy, NY.

42 Diekert, G., and Ritter, M.K.J., 1982, *J. Bacteriol.* **151** 1043.

43 Stoecker, J., and Pope, D., 1986, *Materials Performance* **25**(6) 51.

44 Kobrin, G., 1986, 'Reflections on microbiologically induced corrosion of

stainless steels', in *Biologically Influenced Corrosion*, Dexter, S.C., ed., National Association of Corrosion Engineers, Houston, TX.

45 Borenstein, S.W., 1991, 'Why does microbiologically influenced corrosion occur preferentially at welds?', *Corrosion/91*, paper no. 286, National Association of Corrosion Engineers, Houston, TX.

46 Borenstein, S.W., and Lindsay, P.B., 1987, 'Microbiologically influenced corrosion failure analyses', *Corrosion/87*, paper no. 381, National Association of Corrosion Engineers, Houston, TX.

47 Kearns, J., and Borenstein, S.W., 1991, 'Microbially influenced corrosion of welded stainless alloys for nuclear power plant service water systems', *Corrosion/91*, paper no. 279, National Association of Corrosion Engineers, Houston, TX.

48 Duquette, D.J., and Ricker, R.E., 1986, 'Electrochemical aspects of microbiologically influenced corrosion', in *Biologically Influenced Corrosion*, Dexter, S.C., ed., National Association of Corrosion Engineers, Houston, TX.

49 Borenstein, S.W., and Lindsay, P.B., 1988, 'Microbiologically influenced corrosion failure analyses', *Materials Performance* **27**(3) 51.

50 Borenstein, S.W., 1988, 'Microbiologically influenced corrosion failures of austenitic stainless steels welds', *Materials Performance* **27**(8) 62.

51 Borenstein, S.W., 1991, 'Microbiologically influenced corrosion of austenitic stainless steels weldments', *Materials Performance* **30**(1) 52.

52 Guthrie, P., 1987, *The Diversity of TVA*, Tennessee Valley Authority, Knoxville, TN.

53 Stein, A., 1987, 'Microbiologically induced corrosion – an overview' presented at Third International Symposium on Environmental Degradation of Materials in Nuclear Power Reactors, Traverse City, MI, 30 August–3 September.

54 Ford, T.E., personal communication.

55 Licina, G.J., 1988, *Sourcebook for Microbiologically Influenced Corrosion in Nuclear Power Plants*, EPRI NP-5580, Electric Power Research Institute, Palo Alto, CA, p. 4-7.

56 Scott, P.J.B., and Davies, M., 1989, *Materials Performance* **28**(6) 57.

57 Videla, H., and DeMele, M.F.L. 1987, 'Microfouling of several metal surfaces in polluted seawater and its relation with corrosion', *Corrosion/87*, paper no. 365, NACE, Houston, TX.

58 Little, B., Wagner, P., and Ray, R., 1992, *Corrosion/92*, paper no. 173, NACE, Houston, TX.

59 Schultz, R.W., 1991, *Materials Performance* **30**(1) 58.

60 ASM Metals Handbook, 1985, *Desk Edition*, ASM International, Metals Park, OH, pp. 6–70.

61 Rosales, B.M., and De Schiapparelli, E.R., 1980, *Materials Performance* **19**(8) 41.

62 Videla, H.A., 1986, 'The action of *Cladosoporium resinae* growth on the electrochemical behavior of aluminum', in *Biologically Influenced Corrosion*, Dexter, S.C., ed., National Association of Corrosion Engineers, Houston, TX.

63 Salvararezza, R.C., De Mele, M.F.L., and Videla, H.A., 1983, 'Mechanisms

of microbial corrosion of aluminium alloys', *Corrosion* **39**(1) 26.

64 Schmitt, C.R., 1986, 'Anomalous microbiological tuberculation and aluminium pitting corrosion – case histories', in *Biologically Influenced Corrosion*, Dexter, S.C., ed., NACE, Houston, TX.

65 Crooker, T.W., 1985, *Stand. News*, ASTM, Vol. 54.

66 NACE, 1984, *Corrosion Basics, An Introduction*, National Association of Corrosion Engineers, Houston, TX, p. 217.

67 Sand, W., Brock, E., and White, D.C., 1984, 'Role of sulfur oxidizing bacteria in the degradation of concrete', *Corrosion/84*, paper no. 96, National Association of Corrosion Engineers, Houston, TX.

68 Van Mechelen, A.C.A., and Polder P.B., 1990, 'Degradation of concrete in sewer environment by biogenic sulfuric acid attack', in *Microbiology in Civil Engineering*, Howsam, P., ed., Proceedings of the Federation of European Microbiological Societies, FEMS Symposium No. 59, E. & F.N. Spon, University Press, Cambridge, UK.

69 Gaudy, A.F., and Gaudy, E.T., 1980, *Microbiology for Environmental Scientists and Engineers*, McGraw-Hill, New York.

70 Hall, G.R., 1989, 'Control of microbiologically induced corrosion of concrete in waste water collection and treatment systems', *Materials Performance* **28**(10) 45.

71 Brock, E., and Sand, W., 1990, in *Microbially Influenced Corrosion and Biodeterioration*, Dowling, N.J., Mittleman, M., and Danko, J., ed., National Association of Corrosion Engineers, Houston, TX.

Detection, diagnosis and monitoring

Detection, diagnosis and monitoring are crucial to correcting MIC failures. Resuming the operation of a system without first determining the root cause of failure leads, in many cases, to a need for costly repairs to fix it yet again.

This chapter dicusses methods for detecting, diagnosing and monitoring corrosion problems related to MIC. These methods include monitoring operating parameters and system changes (such as temperature), as well as using nondestructive inspection techniques and various sampling techniques for water, deposits and metal surfaces to gain microbiological, chemical and metallurgical information.

It is often difficult to obtain a direct cause-and-effect relationship in MIC failures between the type of bacteria present and the corrosion attack. Investigators should take samples with the utmost care, to preserve any evidence of microbiological influence.[1] It is best initially to assume that the observed corrosion resulted from corrosion mechanisms other than MIC. If those preliminary assumptions remain unproven, then consider whether MIC could be the contributing factor.

6.1 The importance of diagnosis

Diagnosis is the first step in determining if the corrosion failure is related to MIC.[1] Accurate diagnosis can determine whether to operate with a system experiencing MIC, to institute a MIC control program or to replace components. An accurate diagnosis should determine the severity of the condition, and whether microorganisms have influenced the condition. This step, while obvious, is very often overlooked, given the budget and schedule requirements of an operating system. Only thorough diagnosis can select the proper mitigation and corrective action plan. For example, treatment for a MIC-related failure may involve adding an oxidizing agent to the system, which could accelerate corrosion at oxygen concentration cells if the diagnosis were inaccurate.[2]

6.2   Assessment

Assessment involves taking the history of the system to determine the significance of many isolated events, including corrosion monitoring activities, repair records, water chemistry records, water treatment, the operating environment and preliminary visual observations.

An investigator must then develop an action plan that includes as the first step taking samples of organisms. As noted above, extreme care must be taken to preserve microbiological activity. Specific training or prior laboratory experience working with cultures or biological systems is strongly recommended. Often, samples are scooped up, put into an empty bottle and simply left on someone's desk for analysis. This method provides poor-quality results, not because the laboratory is incapable, but because the person taking the sample was too poorly trained.

Biological test sampling, water sampling and nondestructive examination should be done before cleanup or repair activities. Cleanup activities add substances such as cleaning agents, or remove substances such as corrosion products and exopolymers. Cleanup before taking samples often makes it impossible to determine the cause of failure.

If water is involved, obtain detailed information, such as water temperature, pH, *Eh*, dissolved gases, oxygen concentration, conductivity, suspended solids and turbidity. Take samples and check the water that is actually present against the available records. See Section 6.5.1 for more information about water quality parameters.

For additional information on laboratory tests, see J.N. Tanis's book, *Procedures of Industrial Water Treatment*, Ltan Inc., Ridgefield Connecticut, 1987, a practical reference to routinely used field methods for the water treatment industry.

Because certain characteristics are associated with MIC, document conditions as they are found using a 35 mm camera. If possible, take colour photographs using a camera equipped with closeup lenses.

Table 6.1 is a suggested list of necessary data for failures relevant to MIC failures, and is similar to the format used in G.J. Licina's *Sourcebook for Microbiologically Influenced Corrosion in Nuclear Power Plants*, EPRI NP-5580, Electric Power Research Institute, Palo Alto, CA, 1988. Licina discusses case histories of MIC with specific details of materials of construction, operating conditions, and so forth.

6.3   Failure analysis

Failure analysis constitutes an after-the-fact review of what went wrong. The failures may occur because of a design flaw, a material substitution error, fabrication problems such as a welding defect, or a multitude of other causes. A failure investigation and analysis determines the root

Table 6.1. Data relevant to MIC failures

- Component/system
- Materials of construction
- Operating history
- Method of discovery
- Water chemistry analysis
- Deposit analysis
- Microbiological analysis
- Metallurgical analysis
- Treatment

Table 6.2. The principal stages of a failure analysis

- Collection of background data
- Visual examination
- Nondestructive examination
- Mechanical testing
- Macroscopic/microscopic examination
- Destructive examination
- Chemical analysis
- Determination of root cause of failure

cause of failure, and, based on that, recommends action to prevent similar failures in the future. The principal stages of a failure analysis are listed in Table 6.2. Determining the root cause of failure requires enough information on all contributory influences.

Water, corrosion products and bacteria should be sampled by trained personnel at the site. See Sections 6.5–6.8 for additional details on analysis. Prior preparations are required before taking samples for determining microbiological growth. For example, samples are often taken into specially prepared containers containing a medium, a special broth capable of sustaining the organism. Collecting microbiological samples into clean containers without any preliminary preparation is often a waste of time and money.

Soil and water samples should be placed in sterile containers and kept cold (such as in an ice chest) but not frozen. Water, corrosion products and microbial biomass samples often are stored in one of three ways:

- With a small amount of source water and a fixing agent such as cacodylate buffered 4% gluteraldehyde.
- In a manner so the microorganisms can be cultured and isolated in the laboratory.
- Transferred to specially prepared test kits containing media for particular organisms, as discussed in Section 6.8.

As mentioned previously, the approach should be to preserve samples. Take samples with care for the failure investigation, so *if* any microbiological influence is involved, the evidence is preserved. It is also best to assume that the observed corrosion was from corrosion mechanisms other than MIC. If the preliminary assumptions about the cause of the corrosion cannot be proven, then consider whether MIC is the contributing factor.

6.3.1   Collecting background data and samples

Collect information relating to the manufacturing, processing and service histories of the failed component, and collect drawings and design requirements. For example, if a valve has corroded, track down the records from the vendor and the manufacturer. Obtain the materials of construction from the purchasing information, as well as the specification. Occasionally, materials are installed incorrectly, the wrong material is installed in error, a subsequent processing step such as heat treatments is performed improperly or omitted, or inappropriate material substitutions are made.

Determine if you are dealing with the original service for the failed component. In the example of the valve above, for instance, you would determine if the valve was originally intended for one service and then moved to another location, simply because it was the correct size.

Determine the service history, including when the component was put into service. Record environmental details, normal operating temperatures, and pressures and cycle conditions, as well as overloads and excursions. Obtain records of the abnormal conditions and events, repair records, overhauls and similar information.

Photograph the failed component. Collect all parts of the failed component and include intact parts, if possible, for comparison.

6.3.2   Visual examination

Leave the failed part alone until the person performing the failure analysis and trained in collecting the samples can examine it. Then examine the failed parts and all fragments with the naked eye before cleaning them. Collect the soil, debris, water, paint and the like as essential information for the analysis. For additional information on inspection techniques, see Section 6.4.

6.3.3   Nondestructive examination

A number of techniques are routinely used to examine failed components. To detect surface cracks and discontinuities, the following techniques are often used:

- Magnetic particle inspection of ferrous metals.
- Liquid penetrant inspection.
- Ultrasonic inspection.
- Eddy current inspection.

To examine internal features, the techniques used are:

- Radiographic inspection.
- Acoustic emission inspection.
- Stress analysis.

Additional information for each technique is given in Section 6.4.

6.3.4   Mechanical testing

Mechanical and physical properties were discussed in Section 3.3.1. Mechanical properties are related to behaviour under load or stress and depend on the properties of the phases and how the phases make up the structure of the alloy. They include tensile strength, yield strength, elongation, toughness, hardness and fatigue strength.

Physical properties are inherent to the material itself, and not affected by processing. They include specific gravity, thermal conductivity, thermal expansion, melting point and modulus of elasticity.

Hardness testing is a simple mechanical test useful for evaluating heat treatment operations, approximating tensile strength and determining work hardening and softening or hardening by overheating. Tensile tests, fatigue tests, fracture toughness tests and impact tests are less commonly performed with failed samples. These tests are useful when the information is necessary to confirm specification requirements, or to determine the effects of surface conditions on properties.

6.3.5   Macroscopic/microscopic examination

A microscopic examination of the failed component is essential to reveal details of metal structures too small to be otherwise visible. The specimen for examination should be representative of the failed section.

Metallography

Macroscopic examination involves observing the component using low power (below 10×) magnification, which provides an overall view of the features, such as general size and the nature of defects (for example, pits).

Additional examinations may be performed using a metallurgical microscope, also called a metalloscope, which uses reflected light and is designed for magnifications from about 50× to 3000×. Metal sections are cut from the failed component, mounted in a special mounting compound, polished to a mirror finish and examined. The specimens are then typically acid-etched to bring out the metallurgical features. At 50× to 250×, most of the relevant metallographic features can be observed, such as inclusions and phases. A good description of metallographic practice applicable to most metals is included in the ASM Metals Handbook, *Desk Edition*, ASM International, Metals Park, OH.

Electron microscopy

Scanning electron microscopy (SEM) uses electron beams, instead of light as in an optical microscope, to scan the surface features of a specimen. SEM can examine both polished specimens or original surfaces, such as fracture surfaces or corroded surfaces. SEM can commonly magnify features from about 100× to 30 000×. It gives high resolution and high depth of field. In addition, attachments such as microprobe and energy dispersive X-ray analysis, which can be used on surfaces while in SEM to determine chemical analysis, are often available.

Transmission electron microscopy (TEM) uses electron beams to examine the features of a very thin specimen (<0.5 mm), so that some of the electrons pass through it. The amount of the beam that is transmitted or diffracted differs, thus producing contrast.

6.3.6 *Chemical analysis*

A good source of information about the many chemical analysis test methods is the Annual Book of ASTM Standards, *Analytical Procedures, for Metals, Ores, and Related Materials*, Vol. 03.05 and 03.06.

Two methods of chemical analyses, wet chemical methods and spectrachemical methods, are primarily used. Wet analysis often uses a small amount of material, such as machined-off chips, such as in ASTM E 350 *Standard Test Methods for Chemical Analysis of Carbon Steel, Low-Alloy Steel, Silicon Electrical Steel, Ingot Iron, and Wrought Iron*. The chips should be prepared in accordance with ASTM E 59, *Standard Practice for Sampling Steel and Iron for the Determination of Chemical Composition*.

The spectrochemical analysis uses the spectrograph; its use is covered by ASTM E 356, *Standard Practices of Describing and Specifying the Spectrograph*. In addition, there are various disciplines of surface analysis such as: Auger electron spectroscopy (AES); X-ray photoelectron spectroscopy (XPS); ion-scattering spectroscopy (ISS); secondary ion mass spectroscopy (SIMS); and energetic ion analysis (EIA).

6.3.7   Determining the root cause of failure

When the evaluation is completed, the pieces should fall into place. The ASM Metals Handbook, *Desk Edition*, gives a good overview of the engineering aspects of failure analysis. The authors recommend carefully examining the test results and documentation. Accepting results that confirm one's diagnosis and discarding those that do not is, unfortunately, a common problem. If the test results are questionable, retest.

6.4   Inspection techniques

Inspections may be very industry-specific. Some plants have a scheduled yearly shutdown, which makes it easier to inspect equipment from the inside. Inspection relies on three methods of examination:[3]

- Visual.
- Via supplemental equipment.
- Via sophisticated analytical techniques.

The inspection should begin with nondestructive examinations. It is imperative to avoid surface contamination until after microbiological samples have been removed for analysis.[4]

Many characteristics of MIC are detected by sight and smell and this information must be recorded promptly. A variety of techniques can be used during the inspection, including:

- Visual examination.
- Photographic documentation.
- Sample preparation (water, nodules, soil, metal removal).
- Properties (temperature, pH, salinity, ion concentration).
- Microbiological testing.
- Nondestructive examination.

6.4.1   Nondestructive examination

The principal stages for a failure analysis include the following techniques for nondestructive examination. The inspector is responsible for witnessing, verifying, inspecting and documenting work performed.[5] The ASM Metals Handbook, *Desk Edition*, discusses nondestructive testing techniques.

6.4.2   Visual inspection

Visual inspection is based on what you can see, as well as on using low powered optical lenses. For example, MIC of austenitic stainless steel includes several visual characteristics:[6]

- Pitting at the low points of tanks of pipe (where there is a likelihood of stagnant water).
- Pitting at the air–water interface.
- Concentric rings of rust-like deposits around each pit.
- Reddish mounds on the pit, referred to as tubercles.
- Small pinholes with large cavities underneath.
- Pitting often occurring at or adjacent to welds.

Document the as-received condition with photographs, using a 35 mm camera and closeup lenses.

6.4.3   Magnetic particle inspection

Magnetic particle (MP) inspection detects surface discontinuities in magnetic materials. It is based on using fine iron powder in the presence of a magnetic field. Magnetization is produced by passing an electric current through the piece, or by placing the metal in a magnetic field using an external source. The lines of force are distorted by a crack, under the influence of a magnetic field, the iron powder accumulates and defines the location of the indication or defect. MP can also detect subsurface defects under certain conditions.

MP is very useful for locating small cracks or surface discontinuities in ferromagnetic materials and can also often detect surface features not completely open to the surface. MP's major disadvantage is that it is applicable to ferromagnetic materials only. It is also limited by thin coverings of paint or plating, which may decrease the technique's sensitivity. Special equipment and very large currents are required for operating the equipment.

6.4.4   Liquid penetrant inspection

Liquid penetrant (LP) inspection works by a penetrating liquid drawn into a discontinuity, such as a crack. The penetrant is held there by capillary action, and the excess is removed from the surface. The technician then applies a powder developer, which draws the penetrant out of the crevices and provides a contrasting colour indication. The indication can be further examined, and a decision made as to whether it is a defect.

Surface cracks, surface porosity, metallic oxides, slag and inclusions at weldments are often commonly examined using liquid penetrant. One advantage of LP inspection is its relative affordability and ease of use. No electronic components are involved. The technique does not depend on ferromagnetism and the arrangement of the discontinuity or indication does not matter. The major limitation of the technique is that imperfections must be open to the surface; LP is not useful for detecting subsurface flaws. Moreover, LP may not be applicable to very rough or porous surfaces.

6.4.5   Ultrasonics inspection

Ultrasonics inspection uses high frequency sound to penetrate the metal, and thereby to detect both surface and subsurface discontinuities. It can be used to determine wall thickness as a quantitative measurement, or as an inspection tool to detect cracks or internal flaws. Technicians typically use ultrasonics inspection to detect cracks, cavities, incomplete weld fusion or incomplete joint preparation, and other discontinuities.

6.4.6   Eddy current inspection

Eddy current inspection uses electromagnetic induction to inspect ferrous and nonferrous alloys. Techniques place a test specimen within the magnetic field of a coil carrying alternating current, which then induces eddy currents within the sample.[7]

Eddy current inspection can be used on materials that conduct electricity, and is most applicable to nonmagnetic materials, such as copper alloys or austenitic stainless steels. It does not require direct electrical contact with the piece being inspected, and is adaptable to high speed inspection, such as condenser tubing.

6.4.7   Radiographic inspection

Radiographic testing (RT) uses either X-rays or gamma rays. When an object is exposed to radiation, the radiation will be, in various measures, absorbed, scattered and transmitted. The transmitted image is then recorded on photographic film.

Radiography shows variations in film density. For example, a pit would appear as an indication on the film. Consider the case of a pit at a weld in austenitic stainless steel that produces a seeping through-wall leak, often termed a weeper. Without destructive testing or observation that the weld was leaking, the weld may appear adequate for service, if based solely on the interpretation of film density of the radiograph. A cross-section of a pit of this kind is illustrated in Fig. 6.1, and a radiograph in Fig. 6.2. Radiography is useful for determining the presence of indications, and is sensitive to 2% of the wall thickness.[4]

Figure 6.3 shows several different pit morphologies associated with MIC in austenitic stainless steel. Figure 6.4 shows how a leak path can be established with only a slight appearance of corrosion on the interior surface and relatively little metal loss. Because of difficulty with interpretation, radiography may be unsuitable for measuring pit depth.

6.1 Pitting initiated at root of weld due to MIC. Base metal 304L, weld metal ER308L. (Pit is labelled X6 in Fig. 6.2.) 5×.

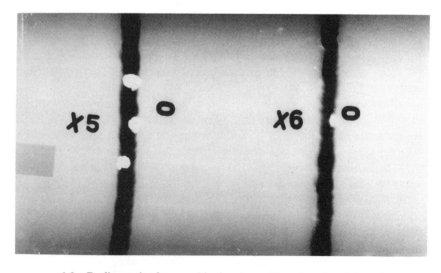

6.2 Radiograph of two welds showing pitting. (reprinted from *Journal American Water Works Association*, Vol. 51, no. 6 (June 1959), by permission. Copyright © 1959, American Water Works Association).

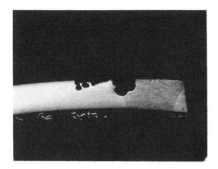
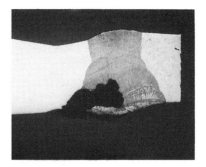
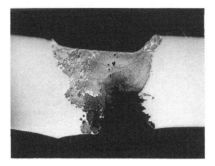

6.3 Pitting austenitic stainless steel weld. Some pitting to base metal. 5×.

Periodic surveillance can measure whether pitting has changed on a system with MIC that has been deemed operational. Nondestructive testing, particularly radiography, can be used to determine the extent of corrosion. For instance, consider the case of through-wall pitting of austenitic stainless steel welds in a large piping system. MIC is suspected. Radiographs may be made of the leaking weld, as well as of the welds nearby. If the welds nearby show evidence of pitting or large subsurface cavities, further investigation and an action plan can be prepared, to deal with the corrosion problem before through-wall leaking occurs.

RT is used extensively on weldments, forgings, mechanical assemblies and castings. It is extremely useful for determining whether internal flaws exist and provides the opportunity to determine if a component is properly placed or fitup of components is adequate. Ferrous and nonferrous alloys are suitable for inspection using RT but it is a very expensive technique compared with all others, in both equipment costs and operator training. Operating costs are relatively high because setup time, in addition to film and processing costs, is considerable.

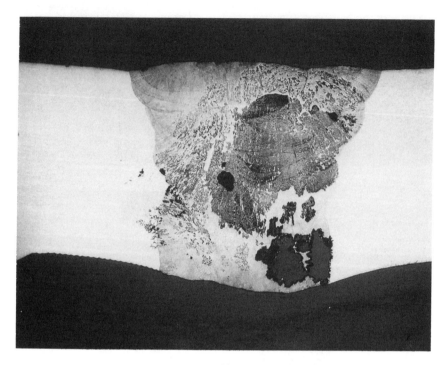

6.4 Pitting of austenitic stainless steel weld due to MIC. Leak path is established with little corrosion on interior surface and little metal loss. 5×.

6.4.8 *Acoustic emission*

Acoustic emission techniques act by receiving high-frequency sound that is emitted by defects, such as a crack under strain. Acoustic emission techniques are very sophisticated in detecting defects and analysing sound signals generated during inspections of materials under stress. The technique is used to locate and determine the significance of discontinuities.

6.4.9 *Stress analysis*

When a load is applied to a material, it is transmitted through to the material as internal force. To find how a material responds to a load is often the objective of courses on the mechanics of materials. For example, such courses may present questions about what materials should be used and what the sizes and proportions of the various elements are, and how they should be constructed to perform a specific function, such as transporting traffic over a river by building a bridge.

Stress is the force per unit area; stress analysis may be described as the study of what happens to stress when a load is applied. The answer may change for different load applications. A static load is a gradually applied load where the members reach equilibrium relatively quickly, such as a truck parked on a bridge. An impact load is a rapidly applied load, such as a truck falling off a trailer on a bridge.

The purpose of stress analysis is usually to find out how the stresses are distributed in a material and a component, and to find out what the effects are. For instance, will it break?

6.5 Water sampling and analysis

Water sampling by various techniques is used to analyse physical and chemical properties, as well as microbiological information.[4,8,9] Water sampling for microorganisms should be done at the site immediately after a failure has occurred. In an ideal situation, the water or aqueous environment is well-characterized before failures occur. See Section 6.8 for details and information on bacteria analysis.

Many laboratories and tests are set up to determine the sanitary quality of water. Bacteria associated with medical problems are generally not the same as those associated with corrosion problems, and may not be relevant. But while the absolute value of a microbial count is useless, the trend is important; for example, whether microbes are either too few or too numerous to count (TNTC). Numbers in between indicate trends, as will repeated tests several months apart or in different seasons, if the environment varies over time. We recommend that technicians use laboratories that specialize in environmental microbiology and water analysis. Laboratories specializing in water chemistry or microbiology, dealing with potable water or pathogenic microorganisms, may be unable to deal effectively with soil or water bacteria associated with MIC.

Water samples can provide information about both the ions in the water and the organisms. Two types of water samples collect organisms: organisms attached to the walls of something are sessile, while organisms floating by in the bulk environment are motile. Roughly 90% are sessile. However, traditionally, technicians take water from flowing streams or systems, immediately upstream and downstream of a problem area, and thus sample primarily motile organisms, which results in far fewer numbers in terms of counting organisms, and correspondingly poorer estimates. The microbiological aspects related to MIC were discussed in Section 2.5.

Additional water samples should be taken for water chemistry analysis. Typical water data required for appraising corrosion and scaling include information presented in Table 6.3. Probably the most significant data

Table 6.3. Typical water chemisty data

- Total dissolved solids
- Total solids
- Calcium
- Magnesium
- Manganese
- Total hardness
- Chlorides
- Sulphates
- Sulphites
- Silica
- Iron
- Oxygen
- Nitrates/nitrites
- Phosphates
- pH
- *Eh*
- Conductivity
- Temperature
- Organic and inorganic carbon

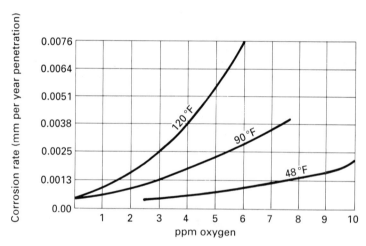

6.5   Effect of oxygen concentration on the corrosion of low carbon steel in tap water at different temperatures (source: 1965, *Materials Performance* **4**(10) 21, reproduced with permission).

from a corrosion perspective is the oxygen dissolved from the air in the water. This relationship of oxygen concentration to corrosion rate is shown in Fig. 6.5.[10]

The collection points for taking water samples can significantly affect the test results. Chemical analysis of the water gives an indication of how

Table 6.4. Water analysis test standards

D 596	Practice for Reporting Results of Analysis of Water
D 887	Practices for Sampling Water-formed Deposits
D 532	Test Method for Iron Bacteria in Water and Water-formed Deposits
D 1125	Test Method for Electrical Conductivity and Resistivity of Water
D 2688	Standard Test Methods for Corrosivity of Water in the Absence of Heat Transfer (Weight Loss Methods)
D 4412	Standard Test Methods for Sulfate-Reducing Bacteria in Water and Water-formed Deposits
D 4778	Standard Test Method for Determination of Corrosion and Fouling Tendency of Cooling Water Under Heat Transfer Conditions

corrosive the water is, as well as whether and how well it can support biological life. A complete water analysis twice yearly may indicate whether the water problem is due only to the corrosiveness of the water, or to microbiological changes.

There are many methods for water analysis. Table 6.4 lists some water analysis test standards relevant to MIC found in American Society for Testing Materials (ASTM), *Water*, Volumes 11.01 and 11.02. This source also gives water analysis methods for salt water, fresh-water, wastewater and many others. For accurate results, pH, dissolved oxygen, temperature and alkalinity should be measured when the sample is taken.

6.5.1 The influence of water quality

The corrosivity of water is influenced by many factors, including dissolved gases, temperature, pH and suspended matter, as well as bacteria, and the effects may be interdependent.[11] We will cover the interactions and the complexities of each.

Dissolved gases

Dissolved oxygen is the primary factor in the corrosion of steels.[11] As discussed in Section 4.6, the role of oxygen is key to most corrosion problems. Corrosion of steel increases when the oxygen content increases and vice versa. The solubility of oxygen in water varies with temperature, pressure and the concentration of ions in the water. Increasing the temperature decreases the oxygen solubility, increasing the partial pressure increases the solubility, while increasing the concentration of ions decreases the oxygen solubility.[11]

For example, open recirculating water systems have a concentration of dissolved oxygen of about 6 ppm.[11] Corrosion of steel in water above a pH of about 6.0 is usually due to the presence of dissolved oxygen, which depolarizes the cathodic reaction and increases the corrosion rate.[12] Neutral water at 20 °C (70 °F) of low salt concentration in the air contains about 8 ppm of dissolved oxygen.[12] The concentration of oxygen decreases as the temperature increases, and as the salt content increases. A figure as low as 0.1 ppm of oxygen can make a difference; it can rapidly increase corrosion rate if it is a dynamic system.[12] For a static system, high concentrations of oxygen may be needed to initiate active corrosion. If the corrosion reaction depletes the oxygen supply and the oxygen is not replaced, the corrosion reaction stops or decreases in rate.

A uniform protective film of magnetite (Fe_3O_4) forms on steel in the presence of water by the reaction:

$$3Fe + 4H_2O \rightarrow Fe_3O_4 + 4H_2 \qquad [6.1]$$

Oxygen can have either a negative or positive effect on the corrosion of steel.[13] If conditions are such that the environment is deaerated, the Fe_3O_4 film isolates the metal surface and forms a protective film, which reduces corrosion. If it is an aerated condition, an uneven, patchy network of corrosion products forms, promoting localized corrosion, by pitting.

Often, oxygen scavengers are added to prevent corrosion, by preventing or reducing the cathodic depolarization process from occurring.

In addition to oxygen, carbon dioxide is a common contributor to corrosion problems. It is more soluble than oxygen in pure water, and readily converts to carbonic acid (H_2CO_3), a weakly acidic acid with a pH of about 6.[13] It promotes corrosion by:

$$CO_2 + H_2O \rightarrow H_2CO_3 \rightarrow H^+ + HCO_3^- \qquad [6.2]$$

The carbonic acid disassociates to form bicarbonate ions (HCO_3^-). If calcium ions are in the water system, they tend to react with the carbonates to form calcium carbonate, which precipitates and forms scales. In addition, ferrous bicarbonate ($Fe(HCO_3)_2$), a soluble salt that results in the thinning of steel, forms from the reaction of carbonic acid (H_2CO_3) and steel.[13]

Chlorine is often added to water to reduce biological growth. Chlorine reacts in water to produce:

$$Cl_2 + H_2O \rightarrow HClO + HCl \qquad [6.3]$$

The chlorine converts to hydrochlorous acid (HClO) and hydrochloric acid (HCl), which tend to reduce the pH.[13] Chlorine can corrode copper alloys by degrading the copper oxide film.

Ammonia (NH_3) is sometimes present in water and aggravates the corrosion of copper alloys.[13]

Temperature

Increased temperatures not only increase corrosion rates, they also decrease the solubility of oxygen in a liquid, thus increasing oxygen's role in corrosion.[13]

In open systems, corrosion can increase up to a maximum solubility of 3 ppm at 80 °C (175 °F).[13] At greater temperatures, the limit is the oxygen content, which reduces the oxygen reduction reaction. Boiling water therefore produces a corrosion reaction similar to that of room temperature water, which has a very high oxygen content.

For closed systems, the reaction is linear with temperature.[13] Other properties, such as viscosity and conductivity, are also affected by temperature. Increasing the temperature increases the diffusion rate of oxygen to a metal surface (i.e. the process whereby molecules intermingle), thus increasing the corrosion rate, since more oxygen is available for the cathodic reaction to occur.

An increase in temperature decreases the solubility of some inorganic salts, such as calcium carbonate ($CaCO_3$) and calcium sulphate ($CaSO_4$). These form deposits at hot spots.[13]

Suspended solids

Suspended solids are particles such as clay, silt, dirt or precipitates, and are common in cooling water systems.[13] Airborne particles are also often drawn into systems, where they add to the solids problems. The particles are often light, soft and nonabrasive, and frequently collect and form barriers at low flowing regions, which adds to problems such as oxygen concentration cells.[13]

The effect of pH

The pH range of a water system should be considered in terms of both the bulk water and local conditions. The actual pH at a metal surface can differ greatly from the bulk water, usually depending on the reactions on the metal surface. The pH is important in governing the rate of corrosion; its influence on the corrosion rate of water containing air depends on the metal being corroded.[14]

Figure 6.6 shows that the corrosion rate for noble metals such as gold and platinum is not affected by pH. Other metals, such as aluminium, zinc and lead, are highly affected in some pH ranges.[14] Severe corrosion occurs at very high and very low pH ranges, because these metals are amphoteric (i.e. metal hydroxides form on the metal surface). The hydroxides are insoluble at approximately neutral pH, but very soluble at high and low pHs. The concrete is an alkaline environment and creates a severely corrosive condition for aluminium.

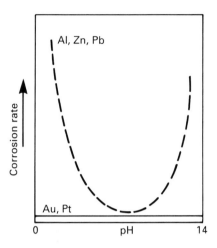

6.6   General curves showing corrosion rates versus pH for aluminium, zinc, lead, gold and platinum (source: Akimov, G.V., 1959, *Corrosion* **15** 455t, reproduced with permission).

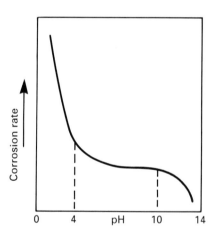

6.7   General curve showing corrosion rate versus pH for iron, nickel, cadmium and magnesium (source Akimov, G.V., 1959, *Corrosion* **15** 455t, reproduced with permission).

For metals such as iron, steel, cadmium and magnesium, another type of corrosion–pH diagram, shown in Fig. 6.7, shows a high rate of corrosion for acid solutions; soluble compounds have formed on the metal surfaces in the region of low pH. These metals form hydroxides that are insoluble in high pH environments and exhibit protective films.

The effect of velocity

Velocity effects are composed primarily of erosion corrosion, cavitation and fretting, as discussed in Section 4.6. The flow of water over a surface influences the corrosion rate.[14] In water with zero velocity, the general corrosion rate is usually very low, while the localized corrosion rate or pitting corrosion is often severe, and the root cause of many failures. In general, some velocity of water results in some form of general corrosion or thinning rather than pitting.

At high velocities, two actions often occur: turbulence and stripping of the protective films. Turbulence often produces nonuniform surface conditions, which may result in corrosion of the bare metal surface.[14]

Typical limiting flow rates are 1.2 m/s (3.9 ft/s) for waters in steel, 1.5 m/s (4.9 ft/s) for copper alloys and 7.5–9 m/s (24.6 ft/s–29.5 ft/s) for stainless steel.[15] Thus, in a process that has a flow of 2.5–3 m/s (8.2 ft/s–9.8 ft/s) for carbon steel, we may predict failure by erosion corrosion. The designer can change materials, reduce the flow or increase the pipe diameter.

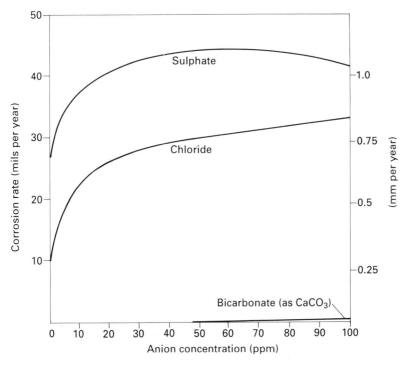

6.8 Influence of sulphate, chloride and bicarbonate on the corrosion of steel (data from Hatch, G.B., and Rice, O., 1959, *Journal American Water Works Assoc.* **51** 719).

Dissolved salts

The corrosivity of water is influenced by dissolved salts in water, which have different degrees of influence at varying concentrations of anions and cations.[16] The hardness ions are calcium, magnesium and HCO_3^- ions which tend to be inhibitive and suppress corrosion.[13] Chloride ions and sulphate ions are not inhibitive but rather promote the rate of corrosion. Water is made less aggressive by increasing the hardness ions. Figure 6.8 shows the relationship of steel in distilled water with the addition of chloride, sulphate and bicarbonate ions. The sulphate is extremely aggressive, the chloride is aggressive, and the bicarbonate tends to inhibit corrosion.[13, 16]

Scale deposits and indices

The factors that affect scale deposition are primarily the total dissolved solids, calcium ion concentration, methyl orange alkalinity and temperature.[17] Calcium and magnesium carbonates have an inverse solubility with temperature; they tend to deposit scales with increasing temperatures.[17]

The tendency for water to deposit scale may be estimated using an equation or index. Two common ones are the Ryzner Stability Index (RSI) and the Langelier Saturation Index (LSI).[18, 19]

LSI is defined as:[20]

$$LSI = pH - pH_s \qquad [6.4]$$

where pH is the actual measured value in the water and pH_s is the pH of saturation calculated from the expression:

$$pH_s = (pK'_2 - pK'_{sp}) + pCa + pAlk \qquad [6.5]$$

where K'_2 is the apparent second dissociation constant of H_2CO_3, K'_{sp} is the apparent solubility product of $CaCO_3$, and pCa is $-\log_{10}$ (total alkalinity) in equivalents per litre.

RSI is defined as:[20]

$$RSI = 2pH_s - pH \qquad [6.6]$$

where pH and pH_s have the same meaning as above.

These numbers are stability indices and provide an indication for a tendency for $CaCO_3$ to form. They do not address a corrosion rate or a rate for the scale to form, nor do they take into account certain factors, such as whether anything in the water would act as an inhibitor. Table 6.5 lists interpretations of both RSI and LSI values giving scale-predicting characteristics.

Langelier and Ryzner stability indices are useful for predicting the

Table 6.5. Scale-predicting characteristics

- If LSI is positive, calcium carbonate tends to form and deposit. This is nonaggressive water.
- If LSI is negative, calcium carbonate tends to dissolve.
  The more negative, the more the water is aggressive.
- If LSI is zero, the water is balanced at calcium carbonate saturation.

- If RSI is below 6.0, calcium carbonate tends to form and deposit. This is nonaggressive water.
- If RSI is greater than 7.0, calcium carbonate tends to dissolve and no scaling occurs. The more above 7.0, the more the water is aggressive.
- If RSI is 6.0, the water is balanced at calcium carbonate saturation.

---

possibility of scale formation, and for indicating water aggressiveness. They are not corrosion indicators.[21] For certain situations, they may not predict whether a protective scale will form, since a salt may form a precipitate only and not a protective scale that covers and tends to protect the metal surface.

Conductivity

Conductivity, or specific conductance, measures the ability to conduct an electrical current. In Section 4.3 we discussed resistivity and conductivity. To review, the electrical resistivity, ρ, is the electrical resistance offered by a material to the flow of current, multiplied by the cross-sectional area of current flow per unit length of current path. Conductivity is the reciprocal to resistivity.[22] Since pure water is highly resistant to current flow, it has a very low conductivity; water containing ions has a higher conductivity. Figure 6.9 shows the relationship of conductivity to corrosion rate.[23]

Section 4.6 covered conductivity and resistivity measurements of soil. Technicians typically make electrical soil resistivity measurements of soil to obtain a relative value for the corrosion data to use in design. The ohm–centimetre is the unit most used for resistivity measurements. The unit 'mho', from ohm spelled backwards, is the common unit for conductivity. The SI unit for conductivity is siemen, in which one mho equals one siemen. The units commonly used are micromhos, or 1 millionth of a mho, or microsiemens. Conductivity meters sometimes read directly in milligrams per litre of dissolved solid, a method that should be used with caution, since the conductivity varies slightly with different waters. Use conductivity data to evaluate water; use soil resistivity data to evaluate soil characteristics; see Table 6.6.

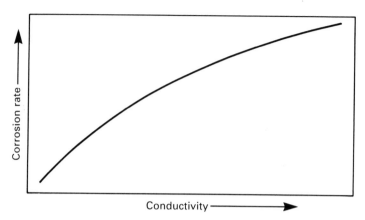

6.9   As conductivity increases, so does the corrosion rate.

Bacteria

As discussed earlier, water often contains microorganisms such as algae, moulds and bacteria, which frequently cause fouling and corrosion problems. Bacteria and other troublesome microorganisms often associated with corrosion and fouling are listed in Table 6.7. Bacteria problems and MIC can usually be avoided by using good quality water, such as water treated by chlorination. As a rule of thumb, water used and then stored for more than three days is no longer of good quality. Many corrosion failures have resulted from initially using good quality water and then, for example, using it as hydrostatic test water in many systems for months at a time. To avoid costly failure, such water should be monitored, retested and retreated.

6.5.2   Types of water

Water has many sources. In determining the root cause of failure in corrosion-related failures, particularly those associated with MIC, the water source is extremely important. (See Section 4.6.4 for more information on water.) Common types of water are listed in Table 6.8. Natural waters include fresh, brackish and salt water, also called seawater.

Fresh water is from rivers, lakes and streams[26] and its salinity or concentration of dissolved salts is low, typically less than 2000 ppm.[26] Fresh water usually contains less than 1000 ppm chloride ion.

Brackish water, found in diluted seawater or tidal rivers, contains roughly 1000–10000 ppm chloride ion, or between 1 and 2.5% sodium chloride. Some areas may have widely varying brackish water, due to seasonal changes.

Table 6.6. The relative values of water corrosivity versus conductivity

Conductivity (micromhos)	Corrosivity
Under 500	Not very corrosive
500–1000	Slightly corrosive
1000–2000	Moderately corrosive
2000–10 000	Corrosive
Over 10 000	Very corrosive

Table 6.7. Some troublesome microorganisms[24, 25]

Slime-forming bacteria
  Flavobacterium
  Mucoids
  Aerobacter
Corrosion-related bacteria
  Desulfovibrio
  Desulfotomaculum
  Desulfomonas
  Clostridium
Iron-depositing bacteria
  Thiobacillus thiooxidans
  Thiobacillus ferrooxidans
  Gallionella
  Sphaerotilus
  Crenothrix
  Pseudomonas
  Sideroscapa
Algae
  Chroococus
  Oscillatoria
  Chlorococcus
Fungi
  Cladosporium
  Aspergillus

Table 6.8. Types of water

- Fresh water
- Brackish water
- Seawater
- Potable water
- Distilled or demineralized water
- Steam condensate
- Boiler feedwater
- Fire protection water/hydrotest water
- Cooling water
- Wastewater

Salt water usually contains 2.5–3.5% sodium chloride. In addition, seawater contains about 65 ppm bromine, 13 strontium and many trace elements.[27]

Connate waters are the waters that remain from ancient seas. They exist in sedimentary and igneous rock, and have been there since the formation of the rock. They sometimes have high salinity because of trapped brackish waters. This type of water is sometimes associated with oil and gas reservoirs.[27]

Potable water (for human consumption) must be free of toxic substances, have low dissolved solids content, and be free of disease-causing organisms. Potable water is often treated with chlorine, sodium hypochlorite or ozone to kill bacteria. A good reference for information about water is *Standard Methods for the Examination of Water and Wastewater*, American Public Health Association, 17th ed., L. Clesceri (ed.), Washington, DC, 1989. Such water is not necessarily free of bacteria, however, and may have bacteria or other organisms that contribute to corrosion problems.

Ostroff discusses bacteria related to potable water and divides them into three groups: natural water bacteria, soil bacteria and bacteria associated with sewage or intestines. Natural water bacteria are not necessarily harmful to people; common bacteria of this type are *Pseudomonas* and *Flavobacterium*. Soil bacteria may appear in water, primarily after heavy rainfall; common bacteria of this type are *Aerbacillus*, *Crenothrix* and *Sphaerotilus*. They are likewise generally not harmful to humans. Bacteria associated with sewage or intestines are pathogenic (i.e. capable of causing disease) to humans; common bacteria of this type are *Clostridium*, *Streptococcus*, *Salmonella* and *Shiggella*.

6.6   Deposit sampling and analysis

Deposits that form on metal surfaces may provide important information. Technicians often use X-ray fluorescence, energy dispersive X-rays (EDS) and wet chemical analysis to analyse corrosion products.

Two other types of common deposits are scale and sludge. Scale forms in place on a surface in contact with the water and sludge usually forms in one place and is deposited in another.[28] Sludge often accumulates where the flow rates are low; for example, when changes in a pipeline have occurred. Such buildup often further reduces flow. Sludges are often less adherent than scales and are typically easier to remove. Scale, as a precipitate, is often very hard and firmly attached to a metal surface, and can also build up on a metal surface and reduce flow. In addition, scale buildup reduces the heat transfer properties in metals.

Chemical analyses of deposits should include measurements of carbon content (organic and inorganic, if possible; if not, then loss-on-ignition) sulphur, chloride, phosphorus, silicon and the major elements in the steel (or base metal). These measurements can determine whether MIC is related to the failure. Licina notes that an organic carbon content of 20% or more indicates likely microbial involvement.[29] He further suggests that manganese and chloride concentrations coupled with high iron contents may indicate the influence of such iron oxidizers as *Gallionella* or *Sideroscapa*. The presence of SRBs may be associated with high sulphur contents, black slimy deposits and the smell of hydrogen sulphide. Deposits are sometimes yellow, and reveal high sulphur content on analysis, which may indicate sulphur oxidizers. See Section 2.2.4 for more information on the general characteristics of microorganisms.

6.7   Destructive sampling and analysis

After sampling water and corrosion products, documenting the conditions with photographs, recording details of visual examination and completing the preliminary examination and nondestructive testing, technicians may need to remove the failed portion of metal for further examination.

Destructive sampling and analysis requires careful recordkeeping, with all records and sketches showing the location of cuts. A log sheet is useful for identifing the location and orientation of the failed section with respect to the direction of flow, top surface, identification numbers, observations (colours, smells), last date of service, date of removal and so forth. A brief description of the background should be included.

Before cutting, keep the surfaces of the failed section as dry as possible. Avoid coolant for the cutting, if possible, and cut well away from the failed portion, to avoid altering the metal structure.

If you are dealing with pitting and suspect subsurface corrosion, especially in austenitic stainless steel welds, you should radiograph the removed piece. The ASME Boiler and Pressure Vessel Code, Section V, Article 2, gives a recommended examination method. Use a pentrameter, as recommended, to ensure film density. Mark the section where the radiographic lead numbers are located, using permanent marker ink or a vibroengraving tool. These marks are needed to locate the pits during metallographic sectioning.

Photograph both the inside and outside surfaces of the removed section. Depending on the conditions of the attack, use higher magnification. If there are light deposits, photograph, lightly brush away the deposits and then photograph again. If there are tightly adhering deposits, you can use light glass bead blasting (69–207 kPa 10–30 psi). Be sure there is no iron contamination in the blast media. Any iron contamination in glass beads will contribute to a surface corrosion problem on the sample under examination. This condition from free iron is termed a rust bloom, and is a problem particularly with austenitic stainless steel.

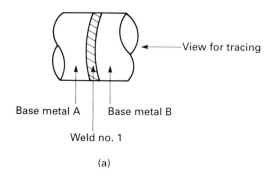

Base metal A   Base metal B

Weld no. 1

(a)

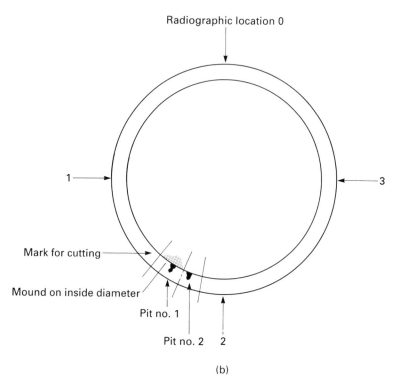

(b)

6.10 Procedure for sampling piping for nondestructive and destructive examination (after Borenstein).

To illustrate destructive examination methods, consider a failure of pitting to an austenitic stainless steel pipe weld. Mark the sections for cutting as shown in Fig. 6.10, using the radiograph to locate the pits. Make a tracing of the pipe and mark the tracing in a similar manner. Give

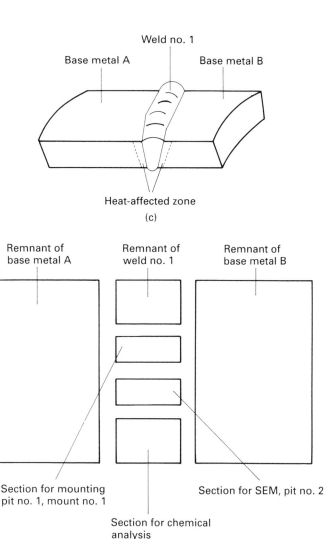

(c)

(d)

6.10 *Continued*

each pit a number and provide additional information about where discoloration is located, if it is a through-wall pit, where the top of the pipe is, the direction of flow and so forth. Identify where sections are to be cut for mounting, for chemical analysis, and for SEM. Plan to cut to one side of the pit, not in the middle of the pit, unless it is quite large. Assign a number and identify each pit on the tracing.

Dry cut (without coolant) sections for metallography, SEM and chemical analysis, as shown in Fig. 6.10, keeping all the pieces attached by a thin section of metal. Photograph the cut sections, including the remnants (Fig. 6.10). Identify remnants and all the sections, so that the original orientation is retained (for example, base metal A and B).

Prepare metallographic mounts. Examine mounts using optical microscopy after polishing, but before etching. Take a photomicrograph at approximately 5× and 100× in order to examine for selective phase attack before etching. Etch, and continue with metallographic examination, taking photomicrographs at 5×, 100× and 500×. Record information about the attack, such as whether the attack is at the weld, the HAZ, the fusion line, in the base metal and so forth.

Sensitization, the susceptibility of stainless steels to intergranular corrosion as a result of chromium carbide precipitation, is usually related to thermal processing. In many cases, it results from welding. (See Section 3.6 for more discussion on sensitization.) Evaluate for sensitization according to ASTM A262 Practice A 'Standard practices for detecting susceptibility to intergranular attack in austenitic stainless steels'. MIC pitting susceptibility of the HAZ appears to be closely related to sensitization.[30]

Examine the exposed pit in the SEM. In many cases involving MIC of austenitic stainless steel weldment, selective attack of one phase of the duplex structure occurs. You can use EDS to identify contaminants in the pits, such as chlorides.

6.8   Biological analysis

Microbiological analysis can determine what organisms are present on a carefully prepared specimen, by several methods.

It is difficult to recommend one method over another. More methods exist, in addition to those common ones discussed below. The identification techniques vary in many ways, including by cost, difficulty and the length of time required to perform the test, and, more important, the methods vary in determining types versus numbers of bacteria. Some tests determine only whether a specific bacteria is present.

You will need to make certain decisions *before* you are able to make samples, such as, for example, what type of microorganism is to be assayed. Thus, if MIC is suspected as contributing to failure, in many cases you will need to prepare the culture media for likely causative organisms before visiting the site. Failure investigators must also be aware that several of the cultured organisms may not be involved in the overall corrosion process, and that some microorganisms that are indeed involved have not been cultured.

As White *et al.* note

> Problems in demonstrating mechanisms of MIC have been complicated by the fact that the classical methods of microbiology that were so successful in the study of infectious diseases (isolation and characterization of the pathogenic species) have proved of little use in the understanding of biofilm dynamics.[31]

He discusses the laboratory analysis for understanding the mechanisms by which bacteria facilitate corrosion; this analysis involves methods for defining the biomass, community structure, nutritional status and metabolic activities of mixed microbial communities.

In addition, it is essential to recognize that there is no direct correlation between the numbers of bacteria detected and the corrosion that can be predicted from their presence.[32] Little and Wagner evaluated standard practices in the United States for quantifying and qualifying sulphate-reducing bacteria in MIC.[33] They note that Pope *et al.* were able to culture high numbers of SRBs from nuclear power plant systems without observing MIC.[34] Pope *et al.* also observed some severe corrosion in the absence of SRBs, but in the presence of acid-producing bacteria.[35]

6.8.1 Media kits

One method of examination relies on selecting certain kinds of microorganisms known to be associated with corrosion. The recovered sample, consisting of corrosion products and microbial biomass, is put directly in media designed for growth of these specific organisms. The microorganisms are then cultured and isolated for identification. These portable examination systems are also called field detection kits. For more information on field testing and kits, see the Gas Research Institute publications, *Microbiologically Influenced Corrosion (MIC): Methods of Detection in the Field* and *Microbiologically Influenced Corrosion (MIC) II: Investigation of Internal MIC and Testing Mitigation Measures*, available from Bioindustrial Technologies Inc., 40200 Industrial Park Circle, Georgetown, TX, 78626. Kits are available commercially.[36–38] In addition to Little's paper, Scott and Davies's research provides an overview of detection methods for SRBs.[33, 39]

Total bacterial counts can also be made using a 'dipstick'. Total bacterial counts are useful with any field testing as a check on the kit for specific organisms. For example, if the dipstick of total count shows a high number of organisms and nothing grows in the kits, the test should be repeated using fresh test kits. The medium may have got too warm or become impaired in some way, resulting in a false positive. As Licina[40] and Pope[41] report, these cultures detect only a fraction of the organisms in the sample (less than 10%).[40] Many sample kits are not organ-specific.[34, 41, 42]

6.8.2   Lipid analysis

Another method of examination relies on identifying microorganisms by their chemical make-up, based on signature lipids content. Several classes of microorganisms have unique signature markers, which may be detected by gas chromotography (GC), or gas chromotography/mass spectroscopy (GC/MS).[43, 44]

6.8.3   Staining

One technique for direct surface staining of bacteria uses a stain, such as acridine orange, to reveal an internal cell enzyme, adenosine. Selected samples are collected, dried, fixed and stained. Bacteria can then be examined under an epifluorescence microscope.

Additional microscopic techniques are used in the laboratory, usually for either specific chemical compounds, such as H_2S, or tailored to specific metabolites of specific microorganisms, such as DNA, RNA or adenosine triphosphate (ATP).

ATP assays estimate the total number of viable organisms. ATP testing is often used with oilfield waters, filtered to remove solids and added to a reagent that releases cell ATP. The enzyme reacts and produces a photo-chemical reaction, and the number of cells can be estimated from the reaction.[33, 45]

Another procedure is the APS method, which is based on the analysis of an internal cell enzyme, adenosine-5'-phosphosulphate (APS) reductase, that is usually in bacteria that reduce sulphate to H_2S.[46] This test is available for field use.

6.8.4   Microscopy

Microscopy often relies on SEM, in which direct observation of micro-organisms on the corrosion product is sometimes possible. With this method a technician prepares the biological material by fixing and de-hydrating, and then sputter-coating it with a thin film of metal. There are disadvantages with the preparations of fixing and drying.

A TEM sample is prepared in a similar manner; the technician then embeds the sample in an epoxy-like material, and sections it to get a thin slice of material for examination. The preparation of the sample often introduces artefacts and affects the interpretation of the surface features, however.

Another method involves using an environmental scanning electron microscope (ESEM), where wet, unfixed biomass can be examined.[47, 48] This technique is discussed in Section 6.8.6.

6.8.5 Mineralogical fingerprints

McNeil *et al.* discuss the technique of using mineralogical fingerprinting as a diagnostic tool for MIC.[49] They used mineralogical data, thermodynamic stability (Pourbaix) diagrams and the simplexity principle for precipitation reactions to rationalize corrosion product mineralogy in a variety of fresh and saline waters, to demonstrate the action of SRBs. They found that if a corrosion process can be shown to have occurred in a pH$-Eh$ range typical of near surface natural environmental conditions, and if no kinetic arguments can be adduced, then mineralogical and geochemical data indicate that the presence of these minerals as corrosion products implies SRB activity. They found that the formation of nonadherent layers of chalcocite (Cu_2S) and the presence of hexagonal chalcocite were indicators of SRB-induced corrosion of copper.[33]

6.8.6 Experimental techniques

A range of experimental techniques can be used to describe the corrosion process of polymer films on metal surfaces, as well as to examine the metal–biofilm interface. Some can be used *in situ* for measuring corrosion rates under biofilms. One method uses Fourier transform infared spectroscopy (FTIR).[50–52] Other experimental techniques for examining the interfacial region (the metal-to-biofilm layer) include X-ray photoelectron spectroscopy (XPS), scanning vibrating probe electrode (SVPE) and energy dispersive X-ray analysis (EDX). Murray *et al.* discuss the techniques and their applications in studying the changes in chemical speciation (the development of species) of the polymer, the formation of metal complexes in the biofilm, and metal distribution between the metal and the surface, the polymer film and the water.

FTIR depends on the measurement of molecular motion by the principles that atoms in a specific chemical environment have characteristic frequencies of motion, such as rotation.[52] Researchers can use the infrared frequency range to analyse a sample and to distinguish changes to the polymer structure. Murray *et al.* discuss FTIR's particular helpfulness in the analysis of principal organic components of biofilm, since the biofilms are composed of proteins and polysaccharides that have characteristic infrared adsorption bands. In addition, time-dependent studies can follow the changing biofilm components over time.

SVPE is a technique that can be used to estimate corrosion at the surface through current density mapping. It maps the current density above a corroding surface without affecting the microbial activity.[52–55] The electrode responds to the varying solution resistances on the surface of a metal. These resistances produce local fields between anodic and cathodic sites. As Murray explains, the vibrating microreference electrode measures the fields produced by the varying current densities, while the probe measures the potential difference between the surface points, finds the electric field from the potential difference and distance between the points, and uses the solution resistivity to relate this to current density.[56] This technique is useful for MIC, because it locates localized corrosion activity with minimum disruption to the biofilm.

The XPS technique is useful for surface analysis of materials, and has been used for studying MIC and the metal–biofilm interaction. During the corrosion process, metal is removed from the surface; it enters the biofilm and goes into the bulk aqueous phase. Murray discusses the work related to MIC using XPS that has focused on the surface phenomena, which occur during the adhesion of bacteria to metal surfaces and the subsequent biofilm formation.[56–61]

EDX uses the action of an electron beam to bombard the surface of a metal sample. X-rays are then emitted. The electrons excite the metal ions and cause ionization. As the ionized ion returns to the ground state, it emits X-rays of a characteristic energy. These characteristic energy peaks identify individual elements. Often, this technique is used with SEM or TEM.[62, 63] The TEM has been used to demonstrate microbiological activity and cells throughout corrosion layers.[64] The ESEM eliminates some of the problems associated with the SEM and disruption of surface features.[65] It accommodates unfixed, uncoated, biological material and provides accurate images of biofilms (Fig. 6.11) without manipulating the sample, which may introduce artefacts.[66]

Wagner *et al.* discuss its use as a tool for documenting the spatial relationships of organisms and their attachment to localized corrosion. Such documentation is essential to establish mechanistic cause-and-effect arguments compared to traditional methods.

6.9   On-line monitoring

On-line monitoring may supplement other monitoring and sampling systems. Together, the monitoring activities may provide an early warning of potential MIC problems.

Monitoring has the potential for abuse, since some operators may consider any information derived from computer systems inherently sacrosanct. Several attitudes are common. For example, consider biocides. A system could be set up to inject biocides automatically if a certain set of parameters should occur. But do we really want a machine telling us when to inject biocides? Operators can forget rules of thumb regarding meters. In a typical scenario, a meter indicates that something is wrong. First, you need to check the meter. Next, you check the instrument that checks the meter, and then check whether the meter could have given you misleading information or got short-circuited. Only then

6.11 Images from ESEM comparing coverage of biofilm when wet and after removal of water (Little, B.J., Ray, R., Pope, R., and Scheetz, R., 1991, 'Biofilms: An ESEM evaluation of artefacts introduced during SEM preparation', *Dev. Ind. Microbiol.* **8**, p. 213). (a) Taken directly from water. (b) After acetone–xylene removal of water.

can you determine if something is genuinely wrong. As you can see from this example, a lot of biocide could have been pumped into a system while an operator trusted that the meter was functioning properly. Maybe it was, and maybe it was not.

6.9.1   Performance monitoring

Routine monitoring of systems may include recording temperatures, pressures and flow rates. MIC of carbon steel is often associated with tubercles forming and causing clogging problems, which reduce flow and often eventually result in through-wall leaks.

As Licina discusses, performance trending may be most useful for detecting MIC effects in carbon steel systems.[67] It cannot alone identify the cause of MIC, or the extent of the degradation resulting from MIC.

6.9.2   Nondestructive examination

Some systems, such as pipe, may be inspected during operation, which is typically done using ultrasonics or radiography.[67–69] Wendell discusses using radiography for inspecting insulated pipes without removing insulation. This method is also useful for rapid detection of wall thinning. (See Section 6.4 for details on nondestructive testing and radiography as a useful tool for MIC examination.)

6.9.3   Corrosion monitoring

On-line corrosion monitoring is often based on exposing coupons to a process stream or a sidestream flow loop. A variation to using coupons is to expose a section of candidate material to the system environment. For example, flanged pipe spools may be installed in a cooling water system in the main flow stream. Electrical resistance and linear polarization resistance techniques are widely used.[70]

The National Association of Corrosion Engineers has a task group (T-3T-3) preparing a document on the various on-line technologies. Perkins reviews categorizing on-line monitoring methods in three groups as shown in Table 6.9. Details about some of these test methods are discussed in Chapter 9.

Perkins notes that group 1 techniques measure the change in the physical geometry of the uncorroded metal, either on a sensing element or directly. Group 2 techniques control either the potential or electrical current density across a metal and conductive fluid interface; and measure whichever is not controlled. Group 3 techniques measure the corrosion byproducts. Perkins give three definitions to clarify the terminology:

Table 6.9. On-line corrosion monitoring methods[70]

Group 1   Methods for measurement of physical metal loss
 1   Corrosion coupons
 2   Electrical resistance probes
 3   Surface activation and gamma radiography
 4   Ultrasonics
 5   Direct electrical methods

Group 2   Methods for measurement of electrochemical properties of corroding surfaces
 1   Linear polarization resistance probes
 2   Galvanic probes (zero resistance ammetry)
 3   Potential measurement probes
 4   Potentiodynamic polarization probes
 5   Electrochemical impedance spectroscopy probes
 6   Electrochemical noise probes

Group 3   Other measurements
 1   Hydrogen probes
 2   Acoustic emission

- On-line is defined as installation of equipment for real-time measurement of metal loss or corrosion rate data in an operating system without disrupting the measurement.
- Real time for corrosion monitoring is defined as measurement taken continuously, or nearly continuously, where the nature of the measurement requires a finite and short cycle time (typically a few minutes to 30 minutes).
- In-line is defined as a measurement made directly in the bulk fluid of the process stream, but requiring extraction for the measurement to be complete.

For MIC monitoring needs, Licina observes that the corrosion response of a system following a perturbation to the system chemistry or temperature may provide a quick indication of microbiological activity in the corrosion process.

Licina *et al.* developed an electrochemical probe designed specifically for on-line monitoring of biofilm activity.[71, 72] The probe uses a two-electrode configuration and intermittent polarization. The probes have been tested in laboratory and plant applications. Microbiological colonies form and measurements of biofilm activity are made. This technique is discussed in Section 9.4.

6.10   Operational measures

Operational measures include keeping a clean, well-maintained system.[73] A clean system is one in which there is adequate water treatment, good water velocity, regular maintenance and cleaning. This aspect of cleaning is a major factor. The objective is to make the attachment of micro-organisms difficult or survival unlikely.

A second consideration is adequate flow. A general rule of thumb is that a flow of 1 m/s (3 feet per second) should be maintained to discourage the growth of microbes.[73] In addition many items discussed in Section 7.3 under the design details about avoiding 'dead' legs, loops and flow paths contribute to reducing or eliminating stagnant or low flow conditions. Johnson provides guidelines for design and operation to minimize stagnation as an operational measure for reducing MIC problems.[74] Some systems are designed to be stagnant. These include duplicate systems or standby systems, and the only flow may be during performance testing or to demonstrate operability, often done as monthly tests. It is essential that these systems get a first-rate water treatment programme to avoid MIC-related problems.

Maintenance is another consideration and is an important part of reducing problems due to MIC. For example, sponge balls in condenser tubes are one method often used as a semicontinuous method for mechanically cleaning water systems.[73] The method is described by Bell, and briefly is where sponge rubber balls a little bit larger than the tubes are used. They are put into the system and collected downstream, collected and reinjected back upstream. It has been effective for some applications to reduce MIC and, when used with chlorination, to reduce corrosion.[75]

6.11   Safety

Safety measures are especially important when technicians are examining corrosion-related failures. All personnel should follow plant safety regulations and US Occupational Safety and Health Administration (OSHA) requirements, especially when entering confined spaces, such as tanks.[76] Use a gas monitor to monitor oxygen and gases present, especially where there is the possibility of gas buildup such as from H_2S or CO_2 from decaying organics or dead fish and plants.

When examining equipment, inspection personnel should minimize the potential for exposure to *Legionella* or other organisms that may adversely affect their health.[77]

Personnel should be prepared for exposure to situations with mist, spray or particulates. Of particular concern with *Legionella* are condenser and cooling towers. An organic vapour mask, as well as gloves, should be worn. After exposure to aerosols, disinfectant should be used to clean the skin.[78-80]

In addition to economics, safety means ensuring no short- and long-term effects to personnel as well as equipment. Dillon notes that the corrosion control method selected must be considered in relation to:[81]

- Fire hazards.
- Explosion hazards.
- Brittle failures.
- Mechanical failures.
- Release of toxic, noxious or hazardous materials.

Suggested reading

1 Characklis, W.G., and Marshall, K.C., 1990, *Biofilms*, John Wiley & Sons, New York.
2 Miller, J.D.A., 1970, *Microbial Aspects of Metallurgy*, American Elsevier, New York.
3 Gaudy, A., and Gaudy, E., 1980, *Microbiology for Environmental Scientists and Engineers*, McGraw-Hill, New York.
4 NACE, TPC 3, 1990, *Microbiologically Influenced Corrosion and Biofouling in Oilfield Equipment*, National Association of Corrosion Engineers, Houston, TX.
5 Dexter, S.C., ed., 1986, *Biologically Influenced Corrosion*, NACE Reference Book No. 8, National Association of Corrosion Engineers, Houston, TX.
6 Licina, G.J., 1989, *Microbial Corrosion: 1988 Workshop Proceedings*, EPRI ER-6345, Electric Power Research Institute, Palo Alto, CA.
7 Guy, A.G., 1976, *Essentials of Materials Science*, McGraw-Hill, New York.
8 NACE, 1984, *Corrosion Basics, An Introduction*, National Association of Corrosion Engineers, Houston, TX.
9 Pope, D.H., 1986, *A Study of Microbiologically Influenced Corrosion in Nuclear Power Plants and a Practical Guide for Countermeasures*, EPRI NP-4582, Electric Power Research Institute, Palo Alto, CA.
10 Licina, G.J., 1988, *Sourcebook for Microbiologically Influenced Corrosion in Nuclear Power Plants*, EPRI NP-5580, Electric Power Research Institute, Palo Alto, CA.
11 Licina, G.J., 1988, *Detection and Control of Microbiologically Influenced Corrosion*, EPRI NP-6815-D, Electric Power Research Institute, Palo Alto, CA.
12 Dillon, C.P., 1986, *Corrosion Control in the Chemical Process Industries*, McGraw-Hill, New York.
13 ASM Metals Handbook, 1989, 9th Ed., Vol. 17, *Nondestructive Evaluation and Quality Control*, ASM International, Metals Park, OH.
14 Mittleman, M., and Geesey, G., 1987, *Biological Fouling of Industrial Water Systems: A Problem Solving Approach*, Water Micro Associates, San Diego, CA.
15 ASM Metals Handbook, 1989, *Desk Edition*, ASM International, Metals Park, OH.

References

1 Licina, G.J., 1988, *Detection and Control of Microbiologically Influenced Corrosion*, EPRI NP-6815-D, Electric Power Research Institute, Palo Alto, CA, p. 4-1.

2   Licina, G.J., 1988, *Detection and Control of Microbiologically Influenced Corrosion*, EPRI NP-6815-D, Electric Power Research Institute, Palo Alto, CA, p. 6-1.

3   Dillon, C.P., 1986, *Corrosion Control in the Chemical Process Industries*, McGraw-Hill, New York, p. 290.

4   Licina, G.J., 1989, *Microbial Corrosion: 1988 Workshop Proceedings*, EPRI ER-6345, Electric Power Research Institute, Palo Alto, CA, p. 6-5.

5   ASM Metals Handbook, 1987, Vol. 13, *Corrosion*, ASM International, Metals Park, OH, p. 417.

6   Licina, G.J., 1989, *Microbial Corrosion: 1988 Workshop Proceedings*, EPRI ER-6345, Electric Power Research Institute, Palo Alto, CA, p. 6-7.

7   Guy, A.G., 1976, *Essentials of Materials Science*, McGraw-Hill, New York, p. 68.

8   APHA, 1985, *Standard Methods for Examination of Water and Wastewater*, 16th ed., American Public Health Association, Washington, DC.

9   Tanis, J.N., 1987, *Procedures of Industrial Water Treatment*, Ltan Inc., Ridgefield, Connecticut.

10 NACE, 1984, *Corrosion Basics, An Introduction*, National Association of Corrosion Engineers, Houston, TX, p. 149.

11 ASM Metals Handbook, 1987, Vol. 13, *Corrosion*, ASM International, Metals Park, OH, p. 489.

12 NACE, 1984, *Corrosion Basics, An Introduction*, National Association of Corrosion Engineers, Houston, TX, p. 131.

13 ASM Metals Handbook, 1987, Vol. 13, *Corrosion*, ASM International, Metals Park, OH, p. 489.

14 Ostroff, A.G., 1979, *Introduction to Oilfield Water Technology*, National Association of Corrosion Engineers, Houston, TX, p. 106.

15 Uhlig, H.H., and Revie, R.W., 1985, *Corrosion and Corrosion Control*, John Wiley & Sons, New York, p. 63.

16 Ostroff, A.G., 1979, *Introduction to Oilfield Water Technology*, National Association of Corrosion Engineers, Houston, TX, p. 104.

17 NACE, 1984, *Corrosion Basics, An Introduction*, National Association of Corrosion Engineers, Houston, TX, p. 150.

18 Ryzner, J.W., 1944, *J. Am. Water Works Assoc.* **36**(4) 1143.

19 Langelier, W.F., 1936, *J. Am. Water Works Assoc.* **28** 1500.

20 ASM Metals Handbook, 1987, Vol. 13, *Corrosion*, ASM International, Metals Park, OH, p. 490.

21 Piron, D.L., 1991, *The Electrochemistry of Corrosion*, National Association of Corrosion Engineers, Houston, TX, p. 178.

22 NACE, 1984, *Corrosion Basics, An Introduction*, National Association of Corrosion Engineers, Houston, TX, p. 191.

23 Betz, 1991, Betz Handbook of Industrial Water Conditioning, Betz Laboratories, Trevose, PA, p. 178.

24 NACE, 1984, *Corrosion Basics, An Introduction*, National Association of Corrosion Engineers, Houston, TX, p. 166.

25 ASM Metals Handbook, 1987, Vol. 13, *Corrosion*, ASM International, Metals Park, OH, p. 118.

26 Ostroff, A.G., 1979, *Introduction to Oilfield Water Technology*, National Association of Corrosion Engineers, Houston, TX, p. 4.

27 Ostroff, A.G., 1979, *Introduction to Oilfield Water Technology*, National Association of Corrosion Engineers, Houston, TX, p. 5.

28 Ostroff, A.G., 1979, *Introduction to Oilfield Water Technology*, National Association of Corrosion Engineers, Houston, TX, p. 55.

29 Licina, G.J., 1988, *Detection and Control of Microbiologically Influenced Corrosion*, EPRI NP-6815-D, Electric Power Research Institute, Palo Alto, CA, p. 6-6.

30 Licina, G.J., 1989, *Microbial Corrosion: 1988 Workshop Proceedings*, EPRI ER-6345, Electric Power Research Institute, Palo Alto, CA, p. 6-15.

31 White, D.C., Nivens, D.E., Nichols, P.N., Mikell, A.T., Kerger, B.D., Henson, J.M., Geesey, G.G., and Clarke, C.K., 1986, 'Role of aerobic bacteria and their extracellular polymers in the facillatation of corrosion: use of Fourier transform infrared spectroscopy and signature phospholipid fatty acid analysis' in Dexter, S.C., ed., *Biologically Influenced Corrosion*, NACE Reference Book No. 8, National Association of Corrosion Engineers, Houston, TX.

32 Soracco, R.J., Pope, D.H., Eggars, J.M., and Effinger, T.N., 1988, *Corrosion/88*, paper no. 83, National Association of Corrosion Engineers, Houston, TX.

33 Little, B., and Wagner, P., 1992, 'Standard practices in the United States for quantifying and qualifying sulfate reducing bacteria in MIC', Proceedings of the Symposium Redefining International Standards and Practices for the Oil and Gas Industry, London.

34 Pope, D.H., 1986, '*A study of MIC in Nuclear, Power Plants and a Practical Guide for Countermeasures*', EPRI NP-4582, Electric Power Research Institute, Palo Alto, CA.

35 Pope, D.H., Duquette, D., Wagner, P., Johannes, A., and Freeman, A., 1989, *Microbiologically Influenced Corrosion: A State-of-the-Art Review*, National Association of Corrosion Engineers, Houston, TX.

36 Korbin, G., 1976, 'Corrosion by microbiological organisms in natural waters', *Materials Performance* **15**(7) 38.

37 Tatnall, R., 1981, 'Case histories: bacteria induced corrosion', *Materials Performance* **20**(8) 41.

38 Stoecker, J.G., 1984, 'Guide for the investigation of microbiologically induced corrosion', *Materials Performance* **23**(8) 48.

39 Scott, P., and Davies, M., 1992, 'Survey of field kits for sulfate reducing bacteria', *Materials Performance* **31**(5) 64.

40 Licina, G.J., 1988, *Sourcebook of Microbiologically Influenced Corrosion in Nuclear Power Plants*, EPRI NP-5580, Electric Power Research Institute, Palo Alto, CA.

41 Pope, D.H., Duquette, D., Wagner, P.C., and Johannes, A.H., 1984, *Microbiologically Influenced Corrosion: A State of the Art Review*, MTI Publication No. 13, Materials Technology Institute of the Chemical Process Industries, St Louis, MO.

42 Tatnall, R.A., Stanton, K.M., and Ebersde, R.C., 1988, 'Methods of testing

for the presence of sulfate reducing bacteria', *Corrosion/88*, paper no. 88, National Association of Corrosion Engineers, Houston, TX.

43   Marr, A.G., and Ingraham, J.L., 1962, 'Effect of temperature on the composition of fatty acids in *Escherichia coli'*, *Journal of Bacteriology* **84** 1260.

44   Drucker, D.B., and Lee S.B., 1981, 'Fatty acid fingerprints of *Streptococci milleri* and *Streptococcus mitis*, and related species', *Int. J. Syst. Bacteriol.* **31** 219.

45   Littman, E.S., 1975, 'Oilfield bacteriacid parameters as measured by ATP analysis', International symposium of Oil Field Chemistry of the Society of Petroleum Engineers of AIME, paper no. 5312, Dallas, TX.

46   EPRI, 1989, *Microbiologically Influenced Corrosion in the Electric Power Generating Industry: A Training Program on Applied Technology*, Electric Power Research Institute, Palo Alto, CA.

47   Little, B.J., Wagner P., and Mansfeld, F. 1991, 'Microbiologically influenced corrosion of metals and alloys', *International Materials Review* **36**(6) 260.

48   Little, B.J., Wagner, P., Ray, R., Pope, R., and Scheetz, R., 1991, 'Biofilms: an ESEM evaluation of artifacts introduced during SEM preparation', *Journal of Industrial Microbiology* **8** 213.

49   McNeil, M.B., Jones, J.M., and Little, B.J., 1991, 'Mineralogical fingerprints for corrosion processes induced by sulfate reducing bacteria', *Corrosion/91*, paper no. 580, National Association of Corrosion Engineers, Houston, TX.

50   Dowling, N.J.E., Stansbury, E.E., White, D.C., Borenstein, S.W., and Danko, J.C., 1989, 'On-line electrochemical monitoring of microbially influenced corrosion', in *Microbial Corrosion: 1988 Workshop Proceedings*, Licina, G.J., ed., EPRI ER-6345, Electric Power Research Institute, Palo Alto, CA.

51   Geesey, G.G., Jolley, J.G., and Hawkins, M.R., 1989, 'Evaluation of occurence and relative concentration of organic products of biofilm development that accumulate at corroding copper surfaces, *Corrosion/89*, National Association of Corrosion Engineers, paper no. 190.

52   Murray, F.E.S., Mitchell, R., and Ford, T.E., 1993, 'Experimental methods for the study of microbially mediated corrosion', *Corrosion/93*, paper no. 295, National Association of Corrosion Engineers, Houston, TX.

53   Issacs, H.S., 1988, *Corrosion Science* **28** 547.

54   Davenport, A.J., Aldykiewicz, A.J., and Issacs, H.G.S., 1992, 'Application of 2D scanning vibrating probe measurements to the study of corrosion inhibitors', *Corrosion/92*, paper no. 234, National Association of Corrosion Engineers, Houston, TX.

55   Franklin, M.J., White, D.C., and Issacs, H.S., 1990, 'The use of current density mapping in the study of microbial influenced corrosion', *Corrosion/90*, paper no. 104, National Association of Corrosion Engineers, Houston, TX.

56   Murray, F.E.S., Mitchell, R., and Ford, T.E., 1993, 'The use of current density mapping in the study of microbial influenced corrosion', *Corrosion/90*, paper no. 104, National Association of Corrosion Engineers, Houston, TX.

57   Andrade, J.D., 1985, *Surface in Interfacial Aspects of Biomedical Polymers*, Vol. 1, Chap. 5, Plenum Press, New York.

58   Magnusson, K.E., and Johanasson, L., 1977, *Studia Biophys* **66** 145.

59   Jolley, J.G., Geesey, G.G., Mankins, M.R., Wright, R.B., and Wichlacz,

P.L., 1988, 'Auger electron spectroscopy and X-ray photoelectron spectroscopy of the biocorrosion of copper by gum arabic, bacteria culture, supernatent and *Pseudomonas atlantica* exopolymer', *Surface and Interface Analysis* **11** 371.

60 Black, J.P., Ford, T.E., and Mitchell, R., (in publication), 'X-ray photoelectron spectroscopy as a technique for the study of metal binding bacterial exopolymers', *Applied and Environmental Microbiology*.

61 Kearns, J.R., Clayton, C.R., Halada, G.P., Gillow, J.P., and Francis, A.J., 1992, 'The application of XPS to the study of MIC', *Corrosion/92*, paper no. 178, National Association of Corrosion Engineers, Houston, TX.

62 Costerston, J.W., and Geesey, G.G., 1986, 'The microbial ecology of surface colonization and of consequent corrosion', in *Biologically Influenced Corrosion*, Dexter, S.C., ed., National Association of Corrosion Engineers, Houston, TX.

63 Characklis, W.G., Little, B.J., and McCaughey, M.S., 1989, 'Biofilms and their effect on local chemistry', *Microbial Corrosion: 1988 Workshop Proceedings*, EPRI ER-6345, Energy Power Research Institute, Palo Alto, CA.

64 Blunn, G., 1986, 'Biological fouling of copper and copper alloys', *Biodeterioration*, VI, Slough, UK.

65 Little, B.J., Ray, R., Pope, R., and Scheetz, R., 1991, 'Biofilms: an evaluation of artifacts introduced during SEM preparation', *Dev. Ind. Microbiol.*

66 Wagner, P.A., Little, B.J., Ray, R.I., and Jones-Meecham, J., 1992, 'Investigation of microbiologically influenced corrosion using environmental scanning electron microscopy', *Corrosion/92*, paper no. 185, Houston, TX.

67 Licina, G.J., 1988, *Detection and Control of Microbiologically Influenced Corrosion*, EPRI NP-6815-D, Electric Power Research Institute, Palo Alto, CA, p. 4-2.

68 Wendell, J., 1987, 'Applications of radiography', *Proceedings: 1987 Seminar on Nuclear Plant Layup and Service Water System Maintenance*, Electric Power Research Institute, Palo Alto, CA.

69 Deardorf, A.F., Copeland, J.F., Poole, A.B., and Rinaca, L.C., 1989, 'Evaluation of structural stability and leakage from pits produced by MIC in stainless steel service lines', *Corrosion/89*, paper no. 514, National Association of Corrosion Engineers, Houston, TX.

70 Perkins, A.J., 1993, 'New developments in on-line corrosion monitoring systems', *Corrosion/93*, paper no. 185, National Association of Corrosion Engineers, Houston, TX.

71 Licina, G.J., Nekoksa, G., 1993, 'On-line monitoring of microbiologically influenced corrosion in power plant environments', *Corrosion/93*, paper no. 403, National Association of Corrosion Engineers, Houston, TX.

72 Licina, G.J., Nekoksa, G., and Howard, R.L., 1992, 'An electrochemical method for on-line monitoring of biofilm activity in cooling water', *Corrosion/92*, paper no. 177, National Association of Corrosion Engineers, Houston, TX.

73 Licina, G.J., 1988, *Sourcebook for Microbiologically Influenced Corrosion in Nuclear Power Plants*, EPRI, NP-5580, Electric Power Research Institute, Palo Alto, CA, p. 3-5.

74 Johnson, C.J., 1988, 'Operational measures for the mitigation of MIC', *Microbial Corrosion: 1988 Workshop Proceedings*, EPRI ER-6345, Electric Power Research Institute, Palo Alto, CA.

75 Bell, R., Lee, R., Tarbin, R., 1985, *State-of-the-art Mechanical Systems for scale and Biofilm Control*, EPRI CS-4339, Electric Power Research Institute, Palo Alto, CA.

76 Code of Federal Regulations, No. 1910.252, *Working in a Confined Space*.

77 Borenstein, S.W., 1989, 'Guidelines for destructive examination of potential MIC-related failures', in *Microbial Corrosion: 1988 Workshop Proceedings*, Licina, G.J., ed., EPRI ER-6345, Electric Power Research Institute, Palo Alto, CA.

78 Edelstein, P.H., 1985, 'Environmental aspects of *Legionella*', *American Society for Microbiology News* **51** 469.

79 Rosa, F., 1986, '*Legionella* and cooling towers', *Heating/Piping/Air Conditioning* **58**(2) 75.

80 Meitz, A., 1986, 'Clean cooling systems minimize *Legionella* exposure' *Heating, Piping, Air Conditioning* **58**(8) 99.

81 Dillon, C.P., 1986, *Corrosion Control in the Chemical Process Industries*, McGraw-Hill, New York, p. 12

Chapter 7

Prevention

Preventing corrosion problems due to MIC is always less expensive than trying to control or to mitigate those problems. The basic tenet of prevention is straightforward: keep microbial growth under control. This chapter provides guidelines for establishing and maintaining a clean system.

7.1   Design

Design is a controversial topic. What constitutes failure? What is the design life? If a piping system is designed for a twenty-year life, is it acceptable for it to fail catastrophically in the twenty-first year? Is it acceptable for it to leak? What if the product is steam, natural gas or fuel oil?

Failures of materials in service usually are traceable to misapplications resulting from:[1]

- Use of the wrong material.
- Improper treatment or fabrication of the material.
- An inadequately controlled (or defined) environment.
- Improper design.

It is not enough for a design to be functional. In selecting the material and corrosion-related portion of a design, the designer needs to consider:

- Cost, both initial and maintenance.
- Safety.
- Consequences of failure.
- Reliability.
- Return on investment.
- The need for inhibitors.
- Unscheduled shutdowns.
- Product loss or contamination.
- Environmental factors.

Designers must consider the following when designing water-containing systems for corrosion prevention due to MIC:[2]

- The water source.
- Its pH.
- Hardness.
- Salinity.
- Dissolved oxygen concentration.
- Impurities.
- Temperature.
- Flow rates.
- Design.

In addition, designers must consider whether to develop an open, closed loop, once-through or recirculating system, since these variables will affect impurity distribution and concentration.

Designers must further consider hydrostatic testing water for all systems, particularly stainless steels. Pitting of stainless steels occurs from the aggressive nature of the chloride ion, which adsorbs on the metal surface, thereby interfering with the formation of the passive oxide film. (Adsorb means to collect a liquid, gas or dissolved substance in a condensed form on a surface.) The small exposed areas that have adsorbed the chloride ion are anodic to the large passive cathode of the oxide film.[3] When corrosion begins, the pH drops, thus accelerating attack and reducing the ability of the oxide film to repair itself.

For buried pipe, the primary concern is corrosion of the outside surfaces. Adequate soil analyses are needed so that the likelihood of corrosion can be adequately predicted. A soil analysis may not be taken as representative of an entire region that covers many kilometres and environments. Tests are applicable only for the areas from which the soil samples were collected. Using data for other locations may result in problems. Factors affecting the corrosiveness of soils are:[4]

- Moisture.
- Alkalinity.
- Acidity.
- Wetness.
- Oxygen content.
- Salts.
- Stray currents, among others.

Soil resistivity is another parameter often reported. Resistivity is the inverse of conductivity, and is discussed in Section 4.3. Uhlig says a poorly conducting soil, whether from lack of moisture, lack of dissolved salts or both, is generally less corrosive than a highly conducting soil.[5]

Additional concerns that should be addressed for buried pipes include low points, areas of stagnation and crevices. These pertain more to internal corrosion problems, but still cause many failures, and are of

Table 7.1. The purposes of coatings[6]

- Corrosion control
- Waterproofing
- Weatherproofing
- Biocide application
- Fireproofing
- Appearance
- Colour coding
- Electrical insulation
- Heat transfer
- Wear resistance
- Safety
- Sanitation/decontamination

---

serious concern. In addition, high points are critical in sewer lines since the corrosion commonly occurs at the air–water interface.

Crevices may also be problems. Weld backing rings commonly used in fitting up welded pipe form a crevice. The use of backing rings may create a series of crevices at weldment that could provide sites for debris and organisms to accumulate, thereby creating ideal conditions for initiating MIC and underdeposit corrosion, as discussed in Section 4.5.

7.1.1 Coatings

Coatings are one method of corrosion prevention. A coating is defined by NACE as a clear or pigmented film-forming liquid that protects the surface to which it is applied from the effects of the environment. A paint is defined as any liquid material containing drying resins and pigments that, when applied to a suitable substrate, will combine with oxygen from the air to form a solid, continuous film over the substrate, thus providing a weather-resistant decorative surface.[6] A coating is considered a protective function, while a paint is considered decorative.

A good quality coating, correctly applied over a properly applied surface, often results in excellent service. However, improperly applied coatings or coatings applied over improperly prepared surfaces are extremely common, and create costly problems. Coatings have many purposes, as shown in Table 7.1.

Protective coatings are usually classified as one of three types, based on their mechanisms of protection; barrier, inhibitive and sacrificial. A barrier coating acts as a barrier between the substrate and the environment. An inhibitive coating, in addition to acting as a barrier, assists in the control of corrosion, typically by the presence of an inhibitor in the pigment. A sacrificial coating corrodes in preference to the structure; for example, zinc is anodic to steel and corrodes preferentially to the steel.

Mechanisms of protection

Coatings offer the following mechanisms of protection:[7]

- Prevent contact between the environment and the substrate (for example, electroplating).
- Restrict or limit contact between the environment and the substrate (for example, organic coatings).
- Release substances that are protective or that attack (for example, chromate primers).
- Produce an electrochemical cell that is protective (galvanizing).

Surface preparation

Most coating systems require adequate surface preparation, which is often a significant part of the overall cost. Lack of preparation or poor quality preparation is a frequent cause of failure.

The leading method for establishing a good surface profile is blasting by abrasive grit. NACE defines four standards for cleaning:[8]

- White metal blast cleaned: produces a grey-white, uniform, metallic colour with no obvious foreign matter present.
- Near-white blast cleaned: foreign matter is removed but differing shades of metallic grey remain.
- Commercial blast cleaned: rust and foreign matter are removed, except for tight specks of oxide or paint.
- Brush-off blast method: light mill scale and tightly adherent rust are allowed.

Other surface preparation methods include:[9]

- Solvent wipe-off.
- Vapour or solution degreasing.
- Steam or hot water cleaning.
- Flame cleaning.
- Mechanical abrasion.
- Electrochemical cleaning.
- Pickling.
- Water blasting.

Materials selection

Coatings may be characterized as follows:[10]

- Primers: the first coatings applied to a surface.
- Secondary or intermediate.
- Topcoats: selected for specific resistance to an environment.

Table 7.2. Factors for selecting painting system[13]

- Regulations
- Service environment
- Substrate
- Service conditions
- Basic function
- Application limitations
- Cost

There are several categories of nonmetallic coating processes, and these produce different effects:

- Phosphate coatings.
- Chromate conversion coatings.
- Painting.
- Rust-preventative compounds.
- Porcelain enamels.
- Ceramic coatings.

Phosphate coatings are made by treating usually iron, steel or aluminium with a dilute solution of phosphoric acid and other chemicals.[11] The surface reacts and is converted to an insoluble protective phosphate layer.

Chromate conversion coatings are made by treating usually iron, steel or aluminium with a solution of chromic acid or chromic salts, hydrofluoric acid or other mineral acids.[12] The chemical attack causes some of the metal's surface to dissolve and form a protective film.

Paint and organic coatings are essentially synonymous. These terms generically describe coatings with an organic base. Often, coatings are based on a film former or binder dissolved in a solvent or water. Selecting a painting system involves several considerations, listed in Table 7.2.

Rust-preventative compounds are removable coatings used for the protection of iron and steel surfaces.[14] These are typically used during shipping, storage and fabrication, or while the surfaces are in use. There are seven general categories of rust-preventative compounds:[14]

- Petrolatum compounds are grease-like and have corrosion inhibitors that provide a barrier film to prevent corrosion.
- Oil compounds are similar to lubricating oils, but have inhibitors to reduce corrosion.
- Hard dry-film compounds form a film after the solvent evaporates, or by a chemical reaction.
- Solvent-cutback petroleum base compounds are residual coatings left after a solvent has evaporated. This term can describe thin films as well as bitumastic hard films.

- Emulsion compounds are emulsion-based, and protect after the water phase has evaporated.
- Water-displacing polar compounds contain wetting agents that coat and protect metal surfaces.
- Fingerprint removers or neutralizers are low viscosity compounds that suppress or neutralize acids and other residues.

The porcelain enamels are glass coatings applied to cast iron, steel or aluminium. They are deemed separate from ceramic coatings, because of their glassy nature.

There are many ceramic coatings; the term typically refers to high temperature coatings based on oxides, carbides, silicides, borides, nitrides, cermets or other inorganic materials.[14] These coatings are usually for protection of wear, oxidation or corrosion, and are often for service at elevated temperatures.

The following summarizes widely used coating materials:[15]

- **Alkyds**: resins prepared by reacting alcohols and acids. They have replaced most oil-based paints and are generally suitable in mild service.
- **Acrylics**: resins prepared by polymerizing acrylic acid and methacrylic acid. They are used in latex paints and have good durability in the atmosphere.
- **Bituminous**: coatings that are coal tar- or asphalt-based resins. They are often used for underground protection and have poor resistance to sunlight and weather.
- **Chlorinated rubber**: a particular film former used as a binder made by chlorinating natural rubber. It is poor in sunlight, but has good chemical and water resistance.
- **Epoxies**: resins made from polyphenols and epichlorhydrin that require catalysts to cure. They are resistant to alkalis, acids and many solvents, and are often used in severe atmospheric services.
- **Epoxy-esters**: modified by adding a drying oil to produce a coating that does not require a catalyst. They are resistant to moisture but do not have good chemical resistance or UV resistance.
- **Latex**: a binder for water-based paint, so a solvent will not be required.
- **Oleoresins**: a combination of oil and resins for slow-drying characteristics.
- **Phenolics**: a family of resins based on phenol-formaldehyde that, when baked, forms a coating with high chemical resistance.
- **Polyesters**: a group of synthetic resins made from reacting dibasic acids with dihydric alcohols.
- **Polyurethanes**: resins usually made by reacting toluene diisocyanate with other materials containing hydroxyl groups, such as polyethers

or polyesters. They offer a wide range of outstanding properties.
- **Vinyls**: a family of synthetic resins based on the polymerization of vinyl compounds, such as vinyl chloride and vinyl acetate. They are often outstanding in marine environments and have good resistance to acids, alkalis and oxidants. They are often used as a topcoat.
- **Zinc, inorganic**: a silicate-based coating with powdered zinc. It is usually topcoated, and is often used in high humidity and salt spray applications. It is poor in acids and alkalis.

See *Corrosion Prevention by Protective Coatings*, by C.G. Munger, NACE, Houston, 1985, for more details about protective coatings.

7.1.2 Cathodic protection

Cathodic protection (CP) is a method of controlling corrosion and is defined as the reduction or elimination of corrosion by making the metal a cathode by means of an impressed direct current or attachment to a sacrificial anode. In terms of the electrochemical behaviour of metals, CP works by polarizing the local cathodes to a more negative potential; the anodes then protect the structure.

There are two methods: making the metal a cathode by means of an impressed DC current, or attaching a different metal to act as a sacrificial anode. Common anodes include magnesium, zinc and aluminium. The CP is commonly used on buried pipe in soil; it protects the outside of the pipe in contact with soil or water (or any electrolyte).

Principles

The principles of CP are based on producing a current flow in the right direction. The principal controlling factor for a metal structure immersed in an electrolyte such as water or soil is the degree of polarization of the cathode.[16] Polarization and Ohm's law were discussed in Section 4.3. Polarization is the shift of potential with current. Ohm's law is the relationship of current (I), voltage (V) and resistance (R)

$$V = IR \qquad [7.1]$$

where V is in volts, I is in amperes and R is in ohms.

Steel in soil typically has a free corroding potential from -0.300 to $-0.700\,V$, versus a copper–copper sulphate reference electrode. This potential is often referred to as pipe-to-soil potential, and is actually the potential measured for the pipe to the half cell reference electrode, usually a copper–copper sulphate electrode for soils or a silver–silver chloride reference electrode for seawater. Table 7.3 gives a comparison

Table 7.3. Standard reference electrodes (repeat of Table 4.4)

Half-cell	Potential at 25 °C (V)
Saturated calomel	+0.242
Normal calomel	+0.280
Tenth normal calomel	+0.334
Silver–silver chloride	+0.222
Copper–copper sulphate	+0.316
Hydrogen	0.000

of other reference electrode potentials with standard hydrogen half-cell reference electrodes.

To illustrate how this comparison works, assume that we measure the half-cell potential of buried steel pipe using a copper–copper sulphate reference electrode and get −0.700 V. To convert to the scale for standard hydrogen electrode, add +0.316 V to get −0.384 V on the scale versus hydrogen. See Section 4.2 about the fundamentals of electrochemistry.

Criteria for CP

There are two criteria for determining whether a metal structure has adequate protection:[17]

- A voltage of −0.85 V versus a saturated copper–copper sulphate reference electrode.
- A minimum negative voltage shift of 100 mV, determined by interrupting the current and measuring the decay.

Both of these criteria are used, but the most common is the −0.85 V versus copper–copper sulphate reference electrode.[17]

The potential versus the environment (the voltage measured relative to the environment, e.g. the soil the metal is embedded in, using a reference electrode) is the first criterion of cathodic protection. It ascertains what electrode potential should be added to a structure to increase its polarization and therefore decrease its corrosion rate.[18]

The corrosion rate decreases when the polarization increases. Examine Fig. 7.1 for the reaction $Fe \rightarrow Fe^{+2} + 2e^{-}$.[19] Piron illustrates how a polarization of 100 mV would increase a structure's life by a factor of 100; thus a polarization of 200 mV appears sufficient, because it corresponds to a potential of −0.700 V versus a saturated copper–copper sulphate reference electrode.[20]

In 1933, J.B. Kuhn proposed using −0.85 V versus a copper–copper sulphate reference electrode for the protection of iron, based on field experiments. His selection became an industry-wide standard. Piron states

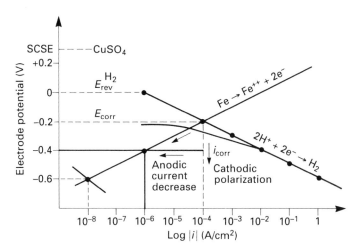

7.1 Schematic of $E \log |i|$ diagram for the corrosion of iron.

that in the presence of sulphate-reducing bacteria, the protection potential should be lowered to $-0.95\,V$. This is generally thought to be the case
by field experience. In poorly drained soils protection potentials of -0.95
are commonly used. In some cases associated with MIC, the $-0.85\,V$ is
insufficient for protection.

NACE RP 01-69 ('Control of external corrosion on underground or
submerged metallic piping systems') gives the criteria for determining
whether an underground piping structure is under cathodic protection.
See A.W. Peabody, *Control of Pipeline Corrosion*, NACE, Houston,
1967, for more details on CP for pipelines.

For many years the negative potential shift of 300 mV was a second
criterion. This criterion was recently removed. To illustrate the problems
associated with it, suppose we immerse iron in a sulphuric acid solution
and get a potential of $-0.570\,V$. A 300 mV shift would produce a potential of $-0.870\,V$ versus a saturated copper–copper sulphate reference
electrode, which is close to the $-0.850\,V$ criterion. The problem is that
the shift potential measures the potential difference between the rest
value and the polarized value, including an *IR* drop. The *IR* drop may
vary from one situation to another, which produces a degree of uncer-
tainty and a different level of protection for the same potential shift
value.[20]

The 100 mV polarization criterion results in a negative polarization of
100 mV from the corrosion potential after the current is interrupted.
Polarization is measured after the current is stopped and before decay,
thus avoiding the *IR* drop problems.

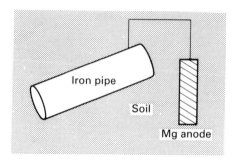

7.2   Cathodic protection with sacrificial anode.

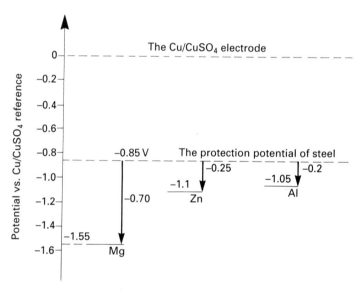

7.3   Comparison of sacrificial anodes (after Piron, D.L., 1991, *The Electrochemistry of Corrosion*, NACE, Houston, TX, reproduced with permission).

Galvanic anode system

Galvanic anodes, also called sacrificial anodes, are simple, using a more active metal to protect a metal structure in a specific environment. For example, Fig. 7.2 shows a magnesium anode protecting a steel structure.

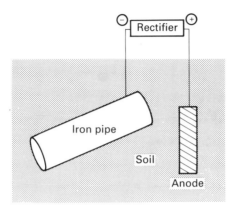

7.4 Cathodic protection with impressed current system.

Magnesium is more active than iron, so it ionizes readily and has a more negative potential than iron in soil or water. The result is a negative polarization of the metal and current flow, which will protect the steel structure. Three metals are typically used as sacrificial anodes in cathodic protection systems: zinc, aluminium and magnesium. Figure 7.3 compares the three metals used for sacrificial anodes.[21] The voltage difference between anode and cathode is typically in the range of 1 volt for galvanic anode systems.[17]

Impressed current system

Impressed current anode systems are more complex. Figure 7.4 shows an impressed current anode system protecting a steel structure, in which there are buried anodes connected to a rectifier and connected to the structure needing protection. The anodes are connected to the positive lead of the DC source.[22] The rectifier changes alternating current to direct current and supplies current to the anodes, which forces them to corrode in preference to the structure. Usually, the impressed current anodes are made from materials that are naturally cathodic to steel. The anode materials have a low consumption rate associated with a high current discharge capability. Impressed current systems often use larger voltage differences between anode and cathode, which produces more efficient current distribution along the protected cathode.

Design

Cathodic protection systems are commonly designed by engineering firms that specialize in corrosion control and cathodic protection design. Several factors affect such design:

- Total current requirement.
- Variations in environment.
- Protective coatings.
- Electrical shielding.
- Economic considerations.
- Metal to be protected.
- Life expectancy or design life.
- Maintenance capabilities.
- Stray current effects.
- Temperature.
- Anode.
- Wire and cable materials.
- Anode backfill, among others.

7.1.3 *The relationship of coatings and cathodic protection to MIC*

Recently, several investigators have examined the role of coatings and cathodic protection to MIC.[23-26] Their research is primarily concerned with external corrosion control using coatings and CP, either alone or in combinations. An important factor appears to be the influence of local soil or electrolyte conditions.

Coatings

Coatings are frequently applied to metal surfaces to protect against corrosion. Coatings are also often applied to external metal surfaces that are buried, such as pipe. The pipe may or may not have CP in addition to the coating, for example. The coatings may be applied to reduce the cost of CP, as well as to prevent soil-side corrosion.

Given the wide variety of coatings available, clear-cut information is difficult to find. All coatings will have small defects such as pinholes, termed holidays, at which corrosion often occurs. The role microorganisms play in the breakdown of coatings on buried pipe is under investigation. Primary factors influencing corrosion at the holidays include the type of coating, its condition, the type of metal substrate, the soil conditions, the exposure time, the temperature and whether CP was applied to the structure.

In general, the porous nature of coatings seems to contribute to their degradation, as a result of microbial processes.[27] Mitton *et al.* describe the degradation of coatings through three stages: in the first stage, the coating behaves as if it were a pure dielectric; in the second stage, water and subsequently dissolved oxygen and ions penetrate through the coating; and in the final stage, the coating separates from the metal substrate.

Cathodic protection

Cathodic protection is frequently used to control corrosion on the external metal surfaces. A cathodic reaction is forced to occur at the surfaces of the metal exposed to soil or electrolyte.[25] The cost of CP is related to the applied current required to maintain the reaction. The amount of current depends on the geometry of the structure, as well as on the composition of the films on the metal surface. Several types of films are common: coatings, corrosion products, calcareous deposits and biofilms.[25] These films influence the current density required to maintain protection using CP. Coatings applied to the external surfaces decrease the cost of CP.

Guezennec and Therene studied the influence of CP on the growth of SRBs and on steel corrosion in marine sediments.[28] They concluded that a protection potential of $-880\,mV$ was not sufficient for protection and that the CP-produced hydrogen promoted the growth of SRBs in the surrounding sediment.[28]

Little et al. showed that cathodically protected stainless steel in artificial seawater can become colonized by aerobic, acid-producing bacteria.[29] Little found that formation of calcareous deposits and initial settlement of microorganisms resulted in a decreased current density being required to maintain a protection potential. Subsequent colonization and pH changes destabilized the calcareous deposits and increased the current density required to maintain the potential.

Little and Wagner found that bacteria can be demonstrated on cathodically protected surfaces.[25] They found that the main consequence of biofilm formation on protected surfaces appeared to be an increase in the current density necessary to polarize the metal to the protection potential. More significant, they found that the presence of a large number of cells on a cathodically protected surface did mean that, in the event that cathodic protection was intermittent, discontinuous or discontinued, the corrosion attack due to the microorganisms would be more aggressive.

7.2   Fabrication

Fabrication is of concern even when the correct materials are specified. Many failures may occur due to improper fabrication or from a lack of attention to details during fabrication.[30]

7.2.1   Startup and layup

Cleanliness in fabrication is essential to prevent MIC; for example, keeping a system clean during fabrication is more important than when it becomes an operating system. There are many references to MIC occurring as the result of fabrication practices or pre-operational testing.[31–35]

Iron contamination of austenitic stainless steel is another common example of corrosion problems related to startup. Any grinding on stainless steel using a tool that worked previously on carbon steel results in embedded particles of carbon steel on the stainless, thus promoting corrosion, often in the form of a rust bloom.

Licina notes that good housekeeping from the beginning of construction to hydrostatic testing and operation is critical to preventing MIC.[36] During construction, drain systems dry. Keep all materials dry and reduce moisture-related problems. Do not store dirty systems and components; if a component gets muddy during construction, for example, procedures should be in place to clean and prepare the component properly for storage or installation.

Carefully monitor and control hydrotest water. Use good quality water, steam condensate or water treated with a biocide for hydrostatic testing. (See Section 8.2 for more details on water treatment.) Improper selection and lack of properly monitored hydrotest water has led to many costly MIC-related failures.

Layup is another problem. Licina discusses various corrosion mechanisms that would be inactive in an operational system, but that appear during layup.[36] For example, water chemistry controls may be more lax. During operations at a power plant, for instance, water purity is carefully monitored to reduce problems associated with impurities in high temperature steam. During layup, however, the conditions may be alternately wet and dry, which would concentrate impurities and result in deposits on metal surfaces that often actively promote corrosion. MIC is prevalent during layup.

Microorganisms may grow in locations with a moist atmosphere; during normal operations, metal surfaces would be subjected to the sheer forces of the water to remove or reduce microbiological growth. Additions of inhibitors or biocides may be helpful. Biocides such as 2 ppm residual chlorine or 0.1 ppm ozone are often recommended for use, taking into account their compatibility with any adverse effect on organics or other construction materials. There are many case histories of MIC failures associated with layup.[33, 36–38]

7.2.2   Residual and fabrication stresses

Fitup is a primary concern during fabrication. Fasteners are one example of this concern for fitup, so choose fasteners that do not introduce stresses.

Residual stresses are also a concern, since they can be introduced by the service conditions. For example, differences in thermal expansion between dissimilar metals can contribute to tensile stresses, compressive stresses or corrosion fatigue.[30] Some theorize that MIC in austenitic stainless steel is associated with residual stresses.[39]

7.2.3 *Surface contamination*

Corrosion products, dirt, impurities, dried seawater and other salts may be present on metal surfaces, and can often contribute to corrosion problems during or after fabrication. They may contribute to MIC by providing nutrients for microbiological growth.

7.3   Maintenance and operation

Maintenance and operational measures include keeping a clean system.[40] A clean water system is one with adequate:

- Water treatment.
- Water velocity.
- Regular maintenance and cleaning.

The objective of cleaning is to make the attachment of microorganisms difficult, or their survival unlikely. This concept of cleanliness is just as important for other chemical or process industries.

A second consideration is adequate flow. Licina gives as a general rule of thumb a flow of 1 m/s (3 feet per second) which should be maintained to discourage the growth of microbes.[40] In addition, many items discussed above regarding design details, such as avoiding 'dead' legs, loops and restricted flow paths, contribute to reducing or eliminating stagnant or low flow conditions. Johnson provides guidelines for design and operation to minimize stagnation as an operational measure for reducing MIC problems.[41] Some systems are designed to be stagnant, such as duplicate or standby systems; the only flow may be during performance testing or to demonstrate operability, often done as monthly tests. It is essential that these systems get a first-rate water treatment program to avoid MIC-related problems.

Maintenance is also crucial in reducing MIC. For example, in Section 6.10 we discussed sponge balls in condenser tubes as one method for mechanically cleaning water systems, a method described by Bell.[40, 42] This method uses sponge rubber balls slightly larger than the tubes. They are put into the system, collected downstream, and reinjected upstream. It is effective for some applications not only to reduce MIC, but also, when used with chlorination, to reduce corrosion.[42]

The guidelines given by Licina for preventing MIC are applicable to most systems:[43]

- Avoid stagnant or low flow conditions.
- Maintain systems by regular cleaning and repair.
- Treat systems when in wet layup conditions, and monitor the treatment if wet conditions continue more than three days.
- Monitor all the water or biocide control treatments to ensure that the systems are functioning properly, and that adequate corrosion protection has been achieved.

7.4   Material selection

Materials are selected to perform specific functions. Unfortunately, however, definitions of corrosive service are varied. As C.P. Dillon states, 'Many corrosion problems are perceived only after the fact; the engineer or maintenance workers in the plant then become immediately concerned with recognizing and understanding the corrosion and metallurgical phenomena, and the economics of repairing, refurbishing, or replacing equipment.'[44]

Ideally, we try to include corrosion control measures in the design phase of a project, and use the best available information for selecting the materials of construction. This selection is usually the area of expertise of the corrosion engineer or materials engineer. Often, these staff members are not available, however, or they may simply not be consulted.

The material selection process includes these steps:

- A review of operating conditions.
- Review of design.
- Selection of candidate materials.
- Evaluation of materials.
- Specification.
- Follow-up monitoring and inspection.

Table 7.4 gives the factors that influence service life from a corrosion standpoint. The most important factors are probably design and materials of construction. The designer and materials engineer must work together to eliminate problems due to design or material failure.

7.4.1   Why failures occur

Material selection can also be examined from an opposing standpoint, by asking 'Why do materials fail?' According to *Corrosion Basics, An Introduction*, failures of materials in service are usually traceable to misapplications resulting from:

- Choice of the wrong material.
- Improper treatment of fabrication of the material.

Table 7.4. Factors that influence service life[45]

- Design
- Materials of construction
- Specification
- Fabrication and quality control
- Operation
- Maintenance
- Environmental conditions

- Inadequately controlled or defined environment.
- Improper design.

7.4.2 Design details

The design details may accelerate corrosion. Design details of primary concern are shape, compatibility, mechanical factors and surfaces.

The shape or geometry of a fabricated component is the most basic element to consider. Determine how to minimize situations that aggravate corrosion. The situation may be innocuous, such as a system that sits stagnant, which may contribute to MIC, as well as to problems with the fluids or solids precipitating out. Crevices and drainage problems, or lack of adequate drainage because of flanges, nozzles and low points below the drain lines, are an invitation to trouble. Crevices should be minimized and 'dead' legs should be avoided. The system may be dynamic, meaning with fluid flow. For example, exceeding recommended velocity may result in damage from mechanisms other than MIC, such as erosion corrosion or cavitation. Other flow-related problems may be associated with splash zones.[46]

Compatibility is a factor, since galvanic corrosion may occur when two dissimilar metals are in direct contact.[47] In addition, unfavourable area effects, such as a small anode located next to a large cathode, should be avoided. See Section 4.3.2 for more on this topic. Dissimilar metals can often be separated by nonmetallic gaskets with insulating sleeves for bolted connections.

Mechanical factors in design include stress (the influence of tensile, compressive or shear stresses), vibration or shock loading, service temperature, fatigue and wear.[48] Stress levels of specific materials with the potential for unexpected catastrophic failure require particular attention.

The surface and the effect of poorly prepared surfaces, rough surfaces and complex shapes may affect a material's performance and contribute to corrosion problems or failures.[48]

Several approaches to control corrosion are discussed by Dillon:[49]

- Change the material of construction.
- Change the environment.
- Put a barrier, such as a coating, between the material and the environment.
- Apply a potential to the metal to be protected.
- Change the design.

7.4.3 MIC environments

MIC problems cannot usually be avoided simply by materials selection. While titanium and other exotic materials appear to be resistant to MIC, they are very expensive and usually impractical for use as materials of fabrication for most components. Moreover, these materials may not be available in many product forms. They are also prone to biofouling to a greater extent that copper alloys, steels or stainless steels.[50, 51]

Always consider the specific environment, temperature and so forth. In general, materials susceptible to crevice corrosion and underdeposit corrosion appear to be susceptible to MIC, while materials resistant to crevice corrosion and underdeposit corrosion appear to be *less* susceptible to MIC.

While, as Licina notes, commonly used materials exhibit various degrees of resistance to MIC, the total consequences of a materials substitution must be evaluated. For example, austenitic stainless steels have greater resistance to MIC than carbon or low alloy steels. The stainless steels rarely have MIC away from weldments and are not prone to plugging problems. For systems in which the water is very corrosive, independent of microbiological activity, consider the following:[50]

- Iron loss to the system from stainless steel will be less than for carbon steel.
- Tubercles involving microorganisms will be less likely to form.
- Blockage and plugging will be reduced.

The alternative view holds that the corrosion of austenitic stainless steel is usually very localized, often producing through-wall pitting and leaks. The corrosion rate or time for pitting to occur and penetrate through-wall is often unpredictable, and stainless steel may not perform any better than carbon steel. If leakage defines failure, a higher alloyed material will not necessarily improve overall performance.

Suggested reading

1 NACE, 1984, *Corrosion Basics, An Introduction*, National Association of Corrosion Engineers, Houston, TX.
2 Licina, G.J., 1988, *Sourcebook for Microbiologically Influenced Corrosion in*

Nuclear Power Plants, EPRI NP-5580, Electric Power Research Institute, Palo Alto, CA.

3 Licina, G.J., 1988, *Detection and Control of Microbiologically Influenced Corrosion*, EPRI NP-6815-D, Electric Power Research Institute, Palo Alto, CA.

4 Dillon, C.P. 1986, *Corrosion Control in the Chemical Process Industries*, McGraw-Hill, New York.

5 Piron, D.L., 1991, *The Electrochemistry of Corrosion*, National Association of Corrosion Engineers, Houston, TX.

6 ASM Metals Handbook, 1987, Vol. 13, *Corrosion*, ASM International, Metals Park, OH.

7 NACE, *Corrosion Data Survey*, 1984, 2 vols., National Association of Corrosion Engineers, Houston, TX.

8 Pludek, V.R., 1977, *Design and Corrosion Control*, Macmillan, New York.

References

1 NACE, 1978, *Corrosion Basics, An Introduction*, National Association of Corrosion Enigneers, Houston TX., p. 338.

2 Licina, G.J., 1988, *Detection and Control of Microbiologically Influenced Corrosion*, EPRI NP-6815-D, Electric Power Research Institute, Palo Alto, p. 3-2.

3 ASM Metals Handbook, 1987, Vol. 13, *Corrosion*, ASM International, Metals Park, OH, p. 490.

4 Fontana, M.G., and Greene, N.D., *Corrosion Engineering*, McGraw-Hill, New York, p. 275.

5 Uhlig, H.H., and Revie, R.W., 1985, *Corrosion and Corrosion Control*, John Wiley & Sons, New York, p. 180.

6 NACE, 1988, *Protective Coating and Lining Course*, National Association of Corrosion Engineers, Houston, TX.

7 NACE, 1984, *Corrosion Basics, An Introduction*, National Association of Corrosion Engineers, Houston, TX, p. 245.

8 NACE, 1984, *Corrosion Basics, An Introduction*, National Association of Corrosion Engineers, Houston, TX, p. 248.

9 NACE, 1984, *Corrosion Basics, An Introduction*, National Association of Corrosion Engineers, Houston, TX, p. 249.

10 NACE, 1984, *Corrosion Basics, An Introduction*, National Association of Corrosion Engineers, Houston, TX, p. 252.

11 ASM Metals Handbook, 1987, Vol. 13, *Corrosion*, ASM International, Metals Park, OH, pp. 29–34.

12 ASM Metals Handbook, 1987, Vol. 13, *Corrosion*, ASM International, Metals Park, OH, pp. 29–35.

13 ASM Metals Handbook, 1987, Vol. 13, *Corrosion*, ASM International, Metals Park, OH, pp. 29–36.

14 ASM Metals Handbook, 1987, Vol. 13, *Corrosion*, ASM International, Metals Park, OH, pp. 29–38.

15 Dillon, C.P., 1986, *Corrosion Control in the Chemical Process Industries*,

McGraw-Hill, New York, p. 272.

16 Dillon, C.P., 1986, *Corrosion Control in the Chemical Process Industries*, McGraw-Hill, New York, p. 281.

17 ASM Metals Handbook, 1987, Vol. 13, *Corrosion*, ASM International, Metals Rark, OH, p. 467.

18 Piron, D.L., 1991, *The Electrochemistry of Corrosion*, National Association of Corrosion Engineers, Houston, TX, p. 218.

19 Piron, D.L., 1991, *The Electrochemistry of Corrosion*, National Association of Corrosion Engineers, Houston, TX, p. 216.

20 Piron, D.L., 1991, *The Electrochemistry of Corrosion*, National Association of Corrosion Engineers, Houston, TX, p. 221.

21 Piron, D.L., 1991, *The Electrochemistry of Corrosion*, National Association of Corrosion Engineers, Houston, TX, p. 223.

22 Dillion, C.P., 1986, *Corrosion Control in the Chemical Process Industries*, McGraw-Hill, New York, p. 283.

23 Paakkonen, S.T., Lockwood, S.F., Pope, D.H., Horner, V.G., Morris, E.A., and Werner, D.P., 1993, 'The role of coatings and cathodic protection in microbiologically influenced corrosion,' *Corrosion/93*, paper no. 293, National Association of Corrosion Engineers, Houston, TX.

24 Videla, H.A., Gomez de Saravia, S.G., de Mele, M.F.L., Hernandez, G., and Hartt, W., 1993, 'The influence of different microbial biofilms on cathodic protection at different temperatures,' *Corrosion/93*, paper no. 298, National Association of Corrosion Engineers, Houston, TX.

25 Little, B.J., and Wagner, P.A., 1993, 'The interrelationship between marine biofouling and cathodic protection,' *Corrosion/93*, paper no. 525, National Association of Corrosion Engineers, Houston, TX.

26 Nekoksa, G., 1989, 'Cathodic protection to control microbiologically influenced corrosion,' in *Microbial Corrosion: 1988 Workshop Proceedings*, Licina, G.J., ed., EPRI ER-6345, Electric Power Research Institute, Palo Alto, CA.

27 Mitton, B., Ford, T.E., LaPointe E., and Mitchell, R., 1993, 'Biodegradation of complex polymeric materials,' *Corrosion/93*, paper no. 296, National Association of Corrosion Engineers, Houston, TX.

28 Guezennec, J., and Therene, M., 1988, 'A study of the influence of cathodic protection on the growth of SRB and corrosion in marine sediments by electrochemical techniques', *First European Federation of Corrosion Workshop on Microbiological Corrosion*, Sintra, Portugal; p. 93.

29 Little, B.J., Wagner, P.A., and Duquette, D., 1988, *Corrosion* **44**(5), 270.

30 NACE, 1984, *Corrosion Basics, An Introduction*, National Association of Corrosion Engineers, Houston, TX, p. 347.

31 Licina, G.J., 1988, *Detection and Control of Microbiologically Influenced Corrosion*, EPRI NP-6815-D, Electric Power Research Institute, Palo Alto, p. 3-4.

32 Borenstein, S.W., and Lindsay, P.B., 1988, 'Microbiologically influenced corrosion failure analyses', *Materials Performance* **27**(54) 51.

33 Kobrin, G., 1976, *Materials Performance* **15**(7) 38.

34 Tatnall, R.E., 1981, *Materials Performance* **19**(8) 41.

35 Stoecker, J.G., 1984, *Materials Performance* **23**(8) 48.

36 Licina, G.J., 1988, *Sourcebook for Microbiologically Influenced Corrosion*, EPRI NP-6815-D, Electric Power Research Institute, Palo Alto, CA, p. 8-3.

37 Tatnall, R.E., 1981, *Materials Performance* **20**(9) 32.

38 Puckorius, P.R., 1983, *Materials Performance* **22**(12).

39 Stein, A.A., 1991, *Proceedings Corrosion/91*, paper no. 107, National Association of Corrosion Engineers, Houston, TX.

40 Licina, G.J., 1988, *Sourcebook for Microbiologically Influenced Corrosion*, EPRI NP-6815-D, Electric Power Research Institute, Palo Alto, CA, p. 3-5

41 Johnson, C.J., 1988, 'Operational measures for the mitigation of MIC', *Microbial Corrosion: 1988 Workshop Proceedings*, EPRI ER-6345, Electric Power Research Institute, Palo Alto, CA.

42 Bell, R., Lee, R., and Korbin, R., 1985, *State-of-the-Art Mechanical Systems for Scale and Biofilm Control*, EPRI CS-4339, Electric Power Research Institute, Palo Alto, CA.

43 Licina, G.J., 1988, *Sourcebook for Microbiologically Influenced Corrosion*, EPRI NP-6815-D, Electric Power Research Institute, Palo Alto, CA, p. 3-7.

44 Dillon, C.P., 1986, *Corrosion Control in the Chemical Process Industries*, McGraw-Hill, New York.

45 ASM Metals Handbook, 1987, Vol. 13, *Corrosion*, ASM International, Metals Park, OH, p. 321.

46 ASM Metals Handbook, 1987, Vol. 13, *Corrosion*, ASM International, Metals Park, OH, p. 350.

47 ASM Metals Handbook, 1987, Vol. 13, *Corrosion*, ASM International, Metals Park, OH, p. 340.

48 ASM Metals Handbook, 1987, Vol. 13, *Corrosion*, ASM International, Metals Park, OH, p. 343.

49 Dillon, C.P., 1986, *Corrosion Control in the Chemical Process Industries*, McGraw-Hill, New York, p. 245.

50 Licina, G.J., 1988, *Sourcebook for Microbiologically Influenced Corrosion*, EPRI NP-6815-D, Electric Power Research Institute, Palo Alto, CA, p. 8-4.

51 Videla, H.A., and DeMele, M.F., 1987, 'Microfouling of several metal surfaces in polluted seawater and its relation with corrosion', *Corrosion/87*, paper no. 365, National Association of Corrosive Engineers, Houston, TX.

Chapter 8

Mitigation

This chapter discusses mitigating corrosion, particularly corrosion due to MIC. Once MIC is diagnosed in a system, a treatment and mitigation programme should be implemented.

Methods of corrosion protection for metallic systems fall into several categories:[1]

- Thermodynamic protection.
- Kinetic protection.
- Barrier protection.
- Structural design.
- Environmental control.
- Metallurgical design.

Thermodynamic protection is based on having a free energy change in the positive direction.[1] For example, rusting of steel is the conversion of metal as iron to the corrosion product, which is rust.[2] The change in energy drives the process; thus the study of thermodynamics. Impressed current cathodic protection is another type of thermodynamic process.

Kinetic protection is based on manipulating the rate of corrosion.[1] The anodic and cathodic halves of the corrosion reaction intersect, and if the rates of either reaction are changed so that the intersection will have a lower current density, the corrosion rate will be reduced.[1]

Barrier protection involves coating the metal with a barrier, which precludes an environment for corroding the substrate. A number of types of barrier coatings are available, including:[1]

- Anodic oxides.
- Ceramic and inorganic coatings.
- Inhibitors.
- Organic coatings.
- Conversion coatings.

Structural design is the source of many corrosion problems where the presence of crevices are a primary culprit. Designs that collect water, and those with poor erosion control, are others.[3] Environmental control calls for closed systems.[1] This method removes a constituent from the corrosion process, such as by adding an oxygen scavenger, which may enhance MIC. Metallurgical design involves choosing the appropriate material for the service condition.[1]

While water treatment is a typical first step, it may not be enough to provide adequate biological control and corrosion inhibition. A biocide will not penetrate tubercles or a dense microbial film on a metal surface. Since inhibitors need contact with the metal surface to be effective, the biofilm often needs to be removed by mechanical or chemical methods.

8.1   Cleaning

Cleaning, or removal of sludge, scale and foulants, is slightly different for MIC problems. To clean a system with MIC, technicians must address how the system has degraded. Deposits such as sludge, scale and debris are common. Sludge is often mud, slime and sediment. Scales are precipitates from dissolved minerals (such as calcium salts), or may be reactants with the corrosion process (such as iron phosphate). They are typically dense and tightly adherent, while sludge is usually mud-like or soft.[4]

Minerals are commonly classified as scale-formers, such as calcium carbonate, magnesium compounds, silicas, iron and manganese. Foulants are commonly deposited as sediment. In addition, fouling deposits may result from the reaction of corrosion products with inhibitors.[4]

The facilitation of cleaning should be built into a system design, such as implementing access ports for pigging, air bumping, chemical injection or inspection ports. Air bumping is air-pressurized to remove loose particles, and is usually used in combination with other methods. Cleaning may be very material-specific. ASTM A380 'Standard recommended practice for cleaning and descaling stainless steel parts, equipment, and systems', for example, is a guide for cleaning stainless steels.

8.1.1   Mechanical cleaning

Factors to be considered in selecting a cleaning process are shown in Table 8.1, and are discussed extensively in the ASM Metals Handbook, *Desk Edition*, ASM International, 1985, Metals Park, OH. Mechanical cleaning is one technique for removing deposits associated with MIC, and typically removes much of the deposits as well as oxides on the metal surfaces. With mechanical cleaning, the contaminant is removed by mechanical means such as wiping, abrasive blasting, brushing, grinding, sanding and chipping. However, it may not effectively remove deposits from pits and crevices. Table 8.2 lists methods used in mechanical

Table 8.1. Factors in selecting a cleaning process

- Identification of material to be removed
- Identification of metal surfaces to be cleaned
- Evaluation of the condition of the surface to be cleaned
- Degree of cleanliness required
- Limitations by the geometry of the structure or component
- Impact of the environment
- Cost

Table 8.2. Mechanical cleaning methods

- Flushing and draining
- Air bumping
- Hydrolasing
- Hydroblasting
- Pigging
- Brushes
- Scrapers
- Water blasting
- Sandblasting
- Shotblasting
- Abrasive blasting

cleaning. Mechanical cleaning is usually performed by contractors who specialize in specific cleaning processes.

The substances on the metal surfaces are often divided into two types: inorganic and organic. Inorganics include mill scale, rust, oxides, corrosion products, abrasive particles, welding flux, paints and other residues. Organics include animal, mineral and vegetable oils, grease, waxes, inhibitors, rust preventatives, some cleaning compounds and other residues.

Abrasive blasting is the primary industrial mechanical method for metal cleaning. Abrasive blast cleaning forces abrasive particles, either dry or suspended in liquid, against the surface to be cleaned.[5] The abrasives used can be metallic grit, metallic shot, sand, glass and many others.

Abrasive blasting is a relatively simple method of cleaning surfaces and is widely used for removing most of the scale and rust. Depending on the surface cleanliness required, blasting may be the initial or only means of cleaning, or may be used in combination with chemical cleaning.

Blasting may use small particles in a stream of air or water to impinge on metal surfaces. This technique in air is called sandblasting, shotblasting or gritblasting, while abrasives in water are termed hydrolasing, hydroblasting, hydrohoning, vapour blasting or honing.

Section 7.1.1 discussed the preparation of metal surfaces for coating. The Steel Structures Painting Council (SSPC) gives surface cleaning specifications that are similar to the Pipe Fabricators Institute (PFI), Standard ES-29, Abrasive Blast Cleaning of Ferritic Piping Materials. The PFI has the following surface finishing levels:

- White metal surface finish (Level 1).
- Near-white metal surface finish (Level 2).
- Commercial surface finish (Level 3).
- Brush-off surface finish (Level 4).

The white metal surface finish is the most extensive and requires that the scale and oxides be completely removed. This can be done on relatively smooth surfaces, but is difficult in pits and crevices. The abrasives for blast cleaning are grit, shot and sand.

Grit is composed of angular metal particles, usually of cast steel or white cast iron shot. Shot is the same as grit, but is spherical. Sand refers not simply to sand but also to other nonmetallics such as garnet, quartz, aluminium oxide, silicon carbide and slag, and is often used when the surface should be protected from metal contamination. For example, austenitic stainless steel should be blasted with iron-free sand, glass beads, stainless steel shot or aluminium oxide, to prevent embedding steel that will immediately rust and break down the passive film on the surface. Hand tools, such as wire brushes, files, grinding disks and the like, for use on stainless steel, should not have been used previously on carbon steel or other alloys.

Pigs, brushes and scrapers are also used to clean the surfaces of pipes or tubes, and are forced through by water or gas pressure. They are most suitable for long straight runs of pipe or tube that do not change direction or section size, and for systems designed for using them with launching and retrieval stations at each end; these may be on-line systems, although the tools can negotiate curves. They are most effective on smooth surfaces and are only partially effective on crevices and pits. If pigs get stuck, usually because of pipes grossly occluded from debris or fouling, the resulting search-and-retrieve missions can be expensive.

8.1.2   Chemical cleaning

Chemical cleaning is often done after mechanical cleaning. Chemical cleaning is characterized by these primary active ingredients:[6]

- Mineral acids.
- Organic acids.
- Chelants.

Chemical cleaning, like mechanical cleaning is usually performed by specialized contractors. As Lutey notes:[7]

When using any of these procedures, it must be emphasized that chemical cleaning in a mitigation situation involves the cleaning of a system or component in which corrosion currently exists. Care must be taken not to create an even greater corrosion condition when cleaning is done.

Chemical cleaning is the most effective method to clean out crevices and pits, and so treats MIC for more than just corrosion problems.

Mineral acids, such as hydrochloric, sulphuric and sulphamic acids, used with inhibitors, are the most commonly used method of chemical cleaning. Inhibitors are used to lower the acid attack on base metal. Phosphoric, nitric and chromic acids are also used in specific applications.

Organic acids, such as formic, acetic and citric, are weak acids, and react to form citrates, acetates and other by-products. Hydrogen builds up under the scale and lifts up the deposits.[8] In addition, the acids bind the dissolved metal ions and carry them away as the surface is being cleaned. The organic acids have an advantage in that they are considered low in their corrosivity to the metal surfaces. Those low corrosion rates can be beneficial if a system needs repeated cleanings, or if the system is incompatible with inhibitors. Moreover, the spent acid is more easily disposed of.

Chelants are cleaning systems that use the metal ion/chelant complex. The chelant binds a metal ion and relies on a pH dependency. Examples of chelants used in cleaning systems are ammoniated citric acid, ethylenediamine tetraacetic acid (EDTA), and n-(hydroxyethyl) ethylenediamine tetraacetic acid (HEDTA). These are effective for removing iron and copper oxides, but ineffective on carbonates, phosphates and hardness deposits.[9] Chelants are also discussed in Section 8.4, under inhibitors.

Acid cleaning should not be used on stainless steel welds unless they have been heat-treated or solution-annealed. The welds (actually the heat-affected zone) may be susceptible to preferential corrosion. See Section 4.6.2 for more details.

8.2   Water treatments

Water is primarily used to transfer or produce steam or cooling water. It is basically corrosive to most metals: contaminants and dissolved gases cause many of the corrosion problems, as discussed in Section 4.6. Factors influencing the corrosion of materials in water systems are listed in Table 8.3.

While water treatment is an important step once MIC has been diagnosed, water treatment alone will not mitigate the problem. Water treatment must be rigorous to provide sufficient biocidal action or corrosion inhibition. Most biocides will not penetrate tubercles. The root cause of failure must be fully addressed; otherwise MIC will return.

Table 8.3. Factors influencing corrosion in water systems[10]

- The physical configuration of the system
- The water chemistry
- The presence of solids in the water
- The presence of dissolved gases
- The flow rate
- The temperature of the water
- The presence of bacteria
- Combinations

8.2.1   Water treatment microbiology

As discussed earlier in Chapter 2, bacteria and other organisms in water and other fluids may cause corrosion, deposits and system degradation. Note that the presence of microorganisms in water does not necessarily indicate a problem. In addition, the quantities of microorganisms present in water do not usually correlate with the extent of damage. In fact, microorganisms may not register as viable in the water testing after a failure has occurred, yet still have been the root cause of failure. So many variables (in bacteria, environments, materials, conditions and so forth) may play a role, that settling on a clear cause-and-effect explanation for failure may be difficult or even impossible. The bacteria may be, and often are, concentrated in deposits.

As discussed, some species of bacteria contribute to corrosion. Bacteria exhibit various growth patterns, depending on their environments. They can often grow in consortia (different species grouped together and supporting growth as a whole). For example, anaerobic and aerobic bacteria may live together in a layered network.

Under the same conditions, different bacteria will exhibit different growth. As Ostroff describes, some do well in warm, moist environments, while others flourish in cooler conditions.[11] Some will lie dormant or in spore form until optimum conditions are available to allow them to grow.[11]

8.2.2   Treatments

Water treatment is a basic subject that can change as chemical companies develop new proprietary compounds for industrial use.[12] A number of different treatment methods are available. The systems constantly change because of varying operating conditions, environmental and regulatory requirements, seasonal changes to the water, and other factors.

Treatment methods depend on water quality analyses. If analyses are not representative, the entire treatment system may be in jeopardy.[13] There are two basic treatment techniques; external or pretreatment, and internal or in-system.[14]

Table 8.4. External water treatment methods

- Aeration
- Clarification
- Filtration
- Precipitation softening
- Ion exchange
- Membrane systems

External treatment

External water treatment methods are listed in Table 8.4 and described below. Aeration mixes air and water, usually with the flows counter to each other. The contact time and air-to-water ratio must be adequate to remove unwanted gases.[15] Aeration is usually used to reduce carbon dioxide, oxidize iron and manganese from well water, and to reduce hydrogen sulphide and ammonia.[15]

Clarification combines three processes for removing suspended matter from raw water. The suspended matter, composed of particles or solids that will not settle out and are often colloidal, is removed by coagulation, flocculation and sedimentation.[16] The coagulation process destabilizes the particles that can be removed, the flocculation process gathers the particles into a larger agglomeration of particles and sedimentation physically removes or settles the particles that were coagulated and flocculated. Filtration is also used to remove solids from water, usually in a mechanical process of physical and chemical adsorption, straining, and additional processing. Filtration does not remove dissolved solids.[17]

Precipitation softening is a group of processes used to reduce hardness, alkalinity, silica and other constitutents.[18] Water is often treated with lime or soda ash for the carbonate ion; this reacts with the water to form an insoluble compound that is precipitated out later.

Ion exchange processes take undesirable ions out and substitute acceptable ions. For example, in sodium zeolite softener, the scale-formers, such as calcium or magnesium ions, are removed and replaced with sodium ions.[19]

Membrane systems are similar to ion exchange processes, except that they use a membrane instead of a chemical. Membrane processes include the following:

- Ultrafiltration (UF).
- Reverse osmosis (RO).
- Electrodialysis (ED).
- Electrodialysis reversal (EDR).

Cooling water

Water is used in cooling systems. The term is generic, and may mean fresh water, seawater or various other types of water. There are three general types of systems: once-through, closed and open. Once-through systems use water and then pipe it back to the source or an outfall – this system may work under considerable constraints, because of ecological concerns. In addition, even increases in temperature may be harmful to certain marine species.[20]

The other two types are recirculated systems. For example, recirculated water is piped through the cooling system, and then cooled by a cooling tower or spray pond.[21] This is an open system. Closed systems feature a contained body of water and often use other water through an exchanger to eliminate the heat built up in the water. Closed systems are often heavily treated and usually kept to relatively small volumes. Open systems are treated, and corrosion control measures are carefully monitored, because scale and dissolved solids build up and concentrate, forming more corrosion problems. An open system removes the heat buildup by evaporation. Ostroff gives the objectives of cooling water treatment as preventing the following:[21]

- The formation of scale on the cooling surfaces.
- Corrosion of metal by the cooling water.
- Fouling of the cooling surfaces and of the cooling tower.
- Deterioration or corrosion of the cooling system materials.

Chemical feed

Chemical feed systems are devices used to add chemicals used in water treatment. They may be classified according to the components used, the material to be fed, the control system used and the application.[22] Several costly problems are associated with poor chemical control systems, including:[22]

- High chemical costs due to overfeeding.
- Inconsistent water quality due to fouling.
- High corrosion rates or plugging.
- High costs due to operation and maintenance personnel.
- Damage to systems or components.

Wastewater

Wastewater is often discarded after use but can be recycled.[23] It must be treated before it is discharged from a plant or system. The water consumption of a plant may be reduced by good water treatment, recycling

Table 8.5. Pollutants[24]

- Organic compounds
- Nutrients
- Solids
- Acids and alkalines
- Metals

---

and reuse of water. Discharge permits issued by the US National Pollutant Discharge Elimination System (NPDES) regulate the amount of pollutants that can be returned to the water source.[24] Commonly encountered pollutants are listed in Table 8.5 and described below.

Organic material in water affects its dissolved oxygen level. Organic compounds are commonly measured as chemical oxygen demand (COD) or biochemical oxygen demand (BOD). BOD is the measure of the oxygen consumed by microorganisms as they use the organics.[25] If the organic load causes the dissolved oxygen level to drop and to reduce the oxygen below a critical level, fish and other aquatic life will die. See Chapter 2 for more details on BOD and COD.

Nitrogen and phosphorus help plants and organisms to grow. Some organisms use nitrogen as a food source and consume oxygen, while phosphorus assists in the growth of algae blooms in surface water.

Solids may contribute to many problems. Solids discharged with water may settle and cover organisms, choking populations of organisms and contributing to oxygen-consuming reactions. In addition, solids may produce turbidity problems, by reducing the light and photosynthesis, which kills some organisms.

Changes in pH caused by the discharge of acids or alkalis affect marine life. Some metals are toxic and affect industry, agriculture and municipal water users.

Gas cleaning

As Ostroff has noted, aeration is a process of mass transfer between a liquid and a gas phase.[26] In the treatment of municipal water, for example, aeration is commonly between water and air. In the treatment of oilfield water, however, aeration is between water and carbon dioxide, carbon monoxide, hydrogen sulphide, methane or nitrogen.[26] So, for example, carbon monoxide is used to remove hydrogen sulphide from flood waters, with no reaction between the gas being absorbed and the water. One gas is absorbed and an undesirable gas is released. This concept is variously termed gas cleaning, aeration, gas stripping, or gas absorption and release.

8.3   Biocides

Biocides (also called antimicrobials) are chemical substances that are poisonous to living organisms. While biostats merely retard the growth of living organisms, bacteriastats are chemicals that inhibit or retard the growth of bacteria.[27] Bactericides, as their name implies, are chemicals that kill or otherwise control bacteria.[28–31]

8.3.1   Classifications

Chemicals used to control microorganisms are classified as two families of biocides: oxidizing and nonoxidizing.[32]

Chlorine and bromine are oxidizers, for example. Chlorine is a common biocide that works by its oxidizing power. It is often added as a gas, producing the reaction:

$$Cl_2 + H_2O \rightarrow HOCl + HCl \qquad\qquad [8.1]$$

The water reacts with the chlorine and produces hydrochlorous acid and hydrochloric acid. The HOCl is the active chemical and dissociates as a function of pH:[32]

$$HOCl \rightarrow H^+ + ClO^- \qquad\qquad [8.2]$$

At pH 7.5, equal concentrations exist.[32] Above pH 7.5, ClO^- dominates up to pH 9.5, where there is total ionization, which means that chlorine is less effective in alkaline environments. A rule of thumb is that chlorine is adequate as a biocide for a pH range of 6.5–7.5.[32] Low pH may accelerate corrosion. Chlorination may be continuously applied, typically in the range of 0.1–0.2 mg/l, or intermittently as 'shock' treatment, typically in the range of 0.5–1.0 mg/l.

Chlorine is commonly used for the disinfection and control of microorganisms, and is considered an excellent form of treatment for bacteria. Strict regulations concerning the use of chlorine gas exist, because of safety issues. Other oxygenating biocides for control of microbial growth include:[33]

- Hypochlorites (e.g. sodium hypochlorite, sodium hypochlorite with sodium bromide and calcium hypochlorite).
- Chlorinated or brominated donor molecules (e.g. isocyanurates, trichloro-*s*-triazinetriones and hydanoins).
- Chlorine dioxide.
- Ozone.

The nonoxidizing biocides react with certain cell components or a reaction in the cell. For example, in one mechanism the cell membrane appears damaged, which destroys or inhibits the microbe's growth. In another mechanism, something in the microbe's ability to metabolize or use energy is damaged, which destroys or inhibits the microbe's growth. These conditions differ from oxidizing biocides, which oxidize the surface or interior structure of the cell.

The *Betz Handbook of Industrial Water Conditioning* describes quaternary ammonium compounds (quats) as cationic surface active molecules. Quats use the mechanism through which they damage cell membranes. This condition makes compounds that are normally kept out of the cell, able to penetrate the barrier. In addition, components inside the cell, such as nutrients, can permeate out. The cell cannot grow and eventually dies. The quantity of concentrations used determines whether it is a biocide or a biostat.

Biocides that interfere with metabolism usually interfere with the electron donor or electron acceptor reactions of organisms. Biocides known to inhibit metabolism include:[34]

- Organotins.
- Bis (trichloromethyl) sulphone.
- Methylenebisthiocyanate (MBT).
- *B*-bromo-*B*-nitrostyrene (BNS).
- Dodecylguanidine salts.
- Di-bromonitropropaneamide (DBNDP).
- Gluteraldehydes.
- Carbonates.
- Amines.
- Quaternary ammonium salts.

8.3.2   Biocides used in cooling water systems

Microbiological activity in cooling water is controlled by biocides and deposit control chemicals.[35] Table 8.6 gives a partial list of biocides used in cooling water systems. In addition to chlorine, chlorine dioxide (ClO_2) is another strong oxidizer used in water treatment. This gas stays as ClO_2 in solution and is used in the pulp and paper industry.[36] It is often used in systems with NH_3 or phenols, because of the low reaction with them.

Bromine release compounds are another group of strong oxidizers used in water treatment. Brominated compounds form hypobromous acid (HBrO), and are effective over a broad pH range. In comparison with hypochlorous acid, at pH 7.5, 50% of the HClO is present, while about 90% of HBrO is present. At pH 8.7, 10% of the HClO is present, while about 50% of HBrO is present.[36]

Organo-sulphur compounds inhibit growth by reducing the ability of the organism to metabolize. MBT is a common organosulphur compound used in controlling algae, fungi and bacteria, particularly *Desulfovibrio*.

Table 8.6. Biocides used in cooling water systems[36]

- Chlorine
- Chlorine dioxide
- Bromine
- Organo-bromide (DBNPA)
- Methylene bisthiocyanate
- Isothiazolinone
- Quaternary ammonium salts
- Organo-tin/quaternary ammonium salts
- Glutaraldehyde

Unfortunately, it is pH-sensitive and is not recommended for use in systems with a pH over 8.[36] An alternative compound to consider is isothiazolinone, which is effective in controlling bacteria over a broad pH range.[36]

Ozone, O_3, is an allotrope of oxygen that is an unstable gas, typically generated on-site. It is a very effective oxidizer that has powerful organism-killing properties, but is a potential safety hazard.[37] The US OSHA considers 10 ppb dangerous to life or health and gives clear guidelines for exposure limits.[37]

8.3.3   Macrofouling control

Macrofouling is fouling caused by large organisms such as mussels. Recently, two specific organisms have been a concern in the central United States: the Asiatic clam and zebra mussel have caused tremendous damage since their accidental introduction into the country.

Asiatic clams are freshwater molluscs that prefer warm water but now extend throughout the southern United States up to Minnesota.[38] The clams do not attach to surfaces; rather, the tiny clam larvae pass through intake screens and settle out in low flowing regions, grow and then, once they have grown to large organisms, are swept into condenser or heat exchanger tubes. They plug the tubes, which encourages the accumulation of debris, and contributes to corrosion.

Zebra mussels, found in rivers and lakes of North America and Europe, entered the Great Lakes in the 1980s.[39] These mussels attach to hard surfaces and form large masses of fouling organisms. They also clog piping systems, reducing the effective diameter of the pipe. Cooling systems seem to provide ideal conditions for the growth of both these organisms.

The primary control methods are mechanical cleaning, thermal backwash, screens and strainers, and biocides. Thermal backwash kills organisms with a blast of hot water. A system can often be designed or modified to divert hot water from the outlet back into intake. As it recirculates, the water provides a macrofouling control when used in the range of 40 °C (104 °F) for one-quarter to one hour.[39]

8.4　Corrosion inhibitors

Inhibitors are chemical substances that slow or prevent a chemical reaction and reduce corrosion. Usually only a relatively small amount is added to the bulk environment.[40]

8.4.1　Electrochemistry

Generally, two primary types of inhibitors, anodic and cathodic, are used. Anodic inhibitors increase polarization of the anode, while cathodic inhibitors increase the polarization of the cathode.[41]

A polarization diagram can illustrate how corrosion inhibitors work. Figure 8.1 shows the terms for a freely corroding metal.[42] As noted in *Corrosion Inhibitors*: 'The line E_aD represents the anodic reactions; line E_cD represents the cathodic reactions. The point of intersection is D, which establishes the open circuit potential (OCP), also called E_{corr}, as well as establishing the corrosion current, I_{corr}.'

Figure 8.2 is a schematic diagram showing the relation of metallic corrosion (D), protection (F and G), and inhibition (P).[42] Generally, the immersed metal may be corroding by reactions under anodic control

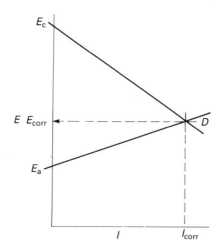

8.1　Anodic and cathodic polarization curves, illustrates significant terms for freely corroding metal (source: NACE, 1973, *Corrosion Inhibitors*, NACE, Houston, TX, reproduced with permission).

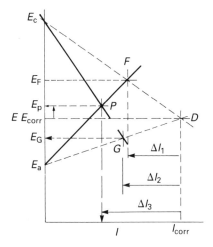

8.2 Schematic diagram showing relation of metallic corrosion, protection and inhibition (source: NACE, 1973, *Corrosion Inhibitors*, NACE, Houston, TX, reproduced with permission).

(E_aF), by use of an anodic inhibitor, under cathodic control (E_cG) by a cathodic inhibitor, or by mixed control $(E_a\text{-}E_cP)$, in which the inhibitor controls both anodic and cathodic reactions. An inhibitor that can control both reactions is clearly more effective. Note that both anodic and cathodic inhibitors reduce the corrosion current, and that the mixed inhibitor reduces the corrosion current most effectively.

Examples of anodic inhibitors are chromates, nitrites, molybdates and phosphates.[43] Examples of cathodic inhibitors are precipitating inhibitors such as calcium bicarbonate, zinc ions, polyphosphate, and phosphonates. Additional types are discussed in various references, such as *Corrosion Basics, An Introduction* and in *Corrosion Inhibitors*.

8.4.2 Oxygen scavengers

Oxygen scavengers remove oxygen by a chemical reaction.[44] Seawater in offshore fields of low salt content contains about 8 ppm dissolved oxygen; the oxygen concentration decreases as the salt content increases and also as the temperature increases.[45] Commonly used oxygen scavengers are:[45, 46]

- Sulphite ion, such as in ammonium bisulphite, sodium sulphite or sodium bisulphite.
- Hydrazine, used in high temperature applications (such as boilers).
- Organic oxygen scavengers.

The factors that determine scavenger effectiveness vary in their impact by applications. Those factors include reaction speed, residence time in the system, operating temperature, operating pressure and pH.[46]

8.4.3   Monitoring inhibitor effectiveness

Inhibitors *must* have access to a clean metal surface to form a barrier to corrosion. Technicians can measure the effect of a corrosion inhibitor in the laboratory by polarization resistance testing since inhibitors function by adsorbing to the metal surface and increasing polarization.[40, 41] In choosing an effective inhibitor, consider the following:[41]

- Can the problems be solved by the addition of a corrosion inhibitor?
- Is the loss from corrosion more expensive than the cost of the inhibitor, including maintenance and operation?
- Is the inhibitor compatible with the process to avoid such adverse effects as foaming?

Inhibitor effectiveness is monitored using various methods, including the following:[47]

- Coupons (standard NACE method).
- Spools.
- Caliper survey.
- Electrochemical methods.

8.4.4   Dispersants

Dispersants are substances that suspend particulate matter by being adsorbed into the surface of particles and giving that surface a high charge.[48] Dispersants are frequently used in deposit and scale control in cooling systems. An electrostatic charge repels similarly charged particles, preventing agglomeration and reducing particle growth.[48] The dispersant also reduces the tendency of particles that precipitate out to bridge together, and in addition the dispersant makes particles more hydrophilic, which renders them less likely to adhere to metal surfaces. This decreases the nutrients in the bulk, and helps to decrease the quantity of biofilm that attaches to the surfaces, by keeping it suspended in the stream.

8.4.5   Surfactants

Surfactants are surface or wetting agents often used to reduce or to prevent fouling of cooling water by insoluble hydrocarbons. They emulsify the hydrocarbon by forming droplets that contain the surfactant. A hydrophobic portion of the surfactant dissolves within the oily portion of the droplet. An electrostatic charge then prevents the hydrophilic groups from coalescing.[48] Used with biocides, surfactants increase the effect in treating MIC by removing organisms that were killed.[49]

8.5   Other methods

Besides maintenance, other methods for mitigation of corrosion due to MIC are to keep adequate flow (typically 1 m/s (3 fps)), thermal treatments, ultraviolet radiation and cathodic protection. Each method must be evaluated to determine the most appropriate choice in terms of economics and an individual situation.

8.6   Replacement of materials

Replacement must be viewed as part of an approach to MIC, albeit generally a last resort.[50] The damage may be so far advanced that the system cannot be repaired. In other cases, the cost analysis may show replacement to be the best economic decision and in still other cases may be unsuitable for water treatment or other corrective actions, unless materials are replaced.

Changing materials is not a panacea. In addition to materials selection, consider fabrication methods, operating procedures, maintenance and water treatment. The source of the problem must be controlled, or degradation will continue. Licina summarizes replacement options as the following:[51]

- Replacement in kind, with increased attention to cleanliness, surface conditions and the like, to improve MIC resistance.
- Replacement with upgraded, more MIC-resistant materials.
- Coatings and linings that improve resistance to MIC of specific components such as piping, heat exchangers, valves and pumps.

8.6.1   Replacement for carbon steels

Replacement of carbon steel is sometimes desired. The most common replacement is substituting stainless steel for carbon steel, which is not without risk. It often only 'buys time', as Licina notes, before MIC is again a problem.[52] The austenitic stainless steels often fail by pitting. Through-wall penetration of 0.6 mm (1/4 inch) thick stainless steel has occurred in as little as six months.[53–55] For some situations, this method may be the right choice. The commonly occurring tuberculation and plugging problems may be unacceptable, while leaks may be acceptable.

8.6.2 Replacement for stainless steels

Replacements for stainless steels often include the 'super' stainless steels, such as higher molybdenum austenitic stainless steels, higher chromium ferritic stainless steels and duplex grade steels.[56] In addition, nickel-based alloys and titanium are often considered as alternatives to stainless steel.

It is unlikely that changing from a normal carbon content (0.08% maximum) austenitic stainless steel to an 'L' grade (0.03% maximum) will significantly change corrosion resistance or affect susceptibility to MIC. Many instances of corrosion due to MIC to Types 304, 304L, 316, 316L grade material are recorded in the literature. It is common to use matching weld filler metals or filler metals with improved corrosion resistance. Thus 304 is usually combined with Type 308 weld filler metal. Type 304L usually is combined with Type 308L weld filler metal. Both base metals may be welded using a higher alloy weld filler metal, such as Type 316 or 316L. Type 316 base metal usually is combined with Type 316 weld filler metal; seldom welded with Type 308 or 308L. My own research indicates that sensitization of the HAZ was not a prerequisite for MIC.[56,57] In addition 'L' grades did not appear more resistant to MIC, either in the weld or HAZ, than standard carbon content austenitic stainless steels. Research is directed at improving the resistance of stainless steel welds to MIC by weld filler metal selection, post-weld heat treatment for weld joints and other methods.

The higher molybdenum austenitic grades include grade 904L with 4.5% molybdenum and 6% compositions, such as Allegheny-Ludlum Steel Corp. product AL-6XN™. Higher molybdenum increases the resistance to pitting in environments known to pit stainless steels, as shown in Fig. 8.3.[58,59] Chlorides in an oxidizing environment are a common example of this environment. The 6% molybdenum austenitic grades have exhibited very good service in seawater and in corrosive environments. Garner studied corrosion of high alloy austenitic stainless steel weldments in oxidizing environments[58] and found that welding has a detrimental effect on pitting resistance.[58] The pitting potential and critical pitting potential were lower for welded than unwelded steel, and the resistance to pitting corrosion increased with molybdenum content.

In general terms, higher alloyed materials with superior resistance to pitting and crevice corrosion in chloride environments would be expected to perform better in MIC environments.

8.6.3 Coatings and linings

Coatings and protective linings may increase corrosion resistance of carbon steels. Coatings and linings may be a replacement in kind for degraded materials, or as a replacement for uncoated or unlined existing materials.

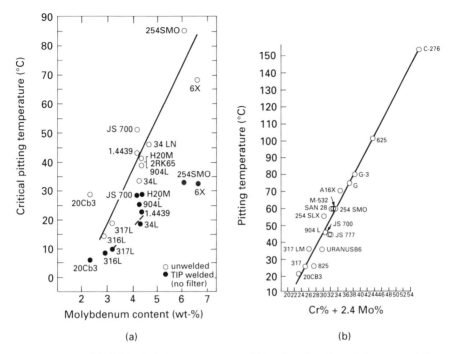

8.3   (a) Critical pitting temperature (directly related to pitting potentials of stainless steels as a function of Mo content (open circles) and in a welded condition (filled circles). (b) Critical temperature for pitting in 4% NaCl + 0.1% $Fe_2(SO_4)_3$ + 0.1 M HCl versus composition for Ni–Cr–Mo–Fe alloys (after Garner, A., 1982, *Materials Performance* **20**(8) 9, reproduced with permission).

Alternatively, the existing materials may have a coating or lining applied in place.[60]

Be cautious, because coatings and linings have been known to fail, with severe consequences. Imperfections in coatings, called holidays, provide sites for corrosion to occur and in addition sheetlike sections may peel away and plug tubes, piping and heat exchangers. The US Nuclear Regulatory Commission has issued an information notice on this type of failure.[61]

Coatings that show good resistance to MIC are being studied. Given the variety of environments the coating may experience, this subject is complex. Literature suggests that many coal-tar-based coatings and synthetic epoxies improve resistance to MIC.[62–67] PVC coatings are reported to fail similarly because of sulphide diffusion through the coating, attacking the surface underneath.[60, 68]

Antifouling coatings may improve resistance to fouling and MIC.[60, 69–71]

These coatings are often termed nontoxic or foul-release coatings. They are based on providing a coating surface with a very low surface tension: the surface is 'slick', resulting in a decreased ability for the fouling organisms to adhere.[70] These coatings reduce macrofouling and may eliminate some MIC problems, because they seem to reduce slime-formers and other microorganisms.

Cement linings may reduce fouling due to microorganisms. The areas near grouted joints and at flaws may present problems. Cement linings may be attacked by sulphur-oxidizing bacteria such as *Thiobacillus*.[60, 72–75] See Section 5.6 about more information related to problems of MIC in concrete.

8.6.4 Material selection criteria

Material selection is discussed in Section 7.4. Construction materials should be selected to keep MIC problems to a minimum. In austenitic stainless steel, the problems generally occur at or near the welds and the HAZ. Frequently, deposits and pitting occur at these locations, so low carbon grade stainless steels with better resistance to pitting corrosion should be used. Weld filler metal should be chosen to make the appro-

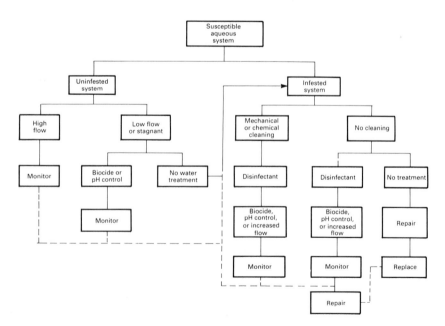

8.4 Control of MIC (after Guthrie, P., 1987, *The Diversity of TVA*, Tennessee Valley Authority, Knoxville, TN).

priate match with the base metal. In addition, cleanliness is essential, so that the area near the weld will be properly prepared, and the corrosion resistance near the weld kept as effective as possible.

Some carbon steels have molybdate inhibition, which reportedly increases general corrosion resistance, lowers pit propagation and reduces crevice attack.

Ideally, a MIC control program should begin before a system is designed and constructed, and well before hydrostatic tests are done and operations begin. Various options to MIC control are given in Fig. 8.4.

Suggested reading

1 NACE, 1984, *Corrosion Basics, An Introduction*, National Association of Corrosion Engineers, Houston, TX.

2 Licina, G.J., 1988, *Sourcebook for Microbiologically Influenced Corrosion in Nuclear Power Plants*, EPRI NP-5580, Electric Power Research Institute, Palo Alto, CA.

3 Licina, G.J., 1988, *Detection and Control of Microbiologically Influenced Corrosion*, EPRI NP-6815-D, Electric Power Research Institute, Palo Alto, CA.

4 Dillon, C.P., 1986, *Corrosion Control in the Chemical Process Industries*, McGraw-Hill, New York.

5 Piron, D.L., 1991, *The Electrochemistry of Corrosion*, National Association of Corrosion Engineers, Houston, TX.

6 ASM Metals Handbook, 1987, Vol. 13, *Corrosion*, ASM International, Metals Park, OH.

7 NACE, 1973, *Corrosion Inhibitors*, National Association of Corrosion Engineers, Houston, TX.

8 Betz, 1991, *Betz Handbook of Industrial Water Conditioning*, Betz Laboratories, Trevose, PA.

9 Licina, G.J., Carney, C., Cubicciotti, D., and Lutey, R., 1990, 'MICPro – predictive software for microbiologically influenced corrosion', *Corrosion '90*, paper no. 536, National Association of Corrosion Engineers, Houston, TX.

10 Conley, E.F., Bell, R.J., Reid, D.J., and Bennett, B.A., 1993, *Service Water Systems Sourcebook*, EPRI ER-6345, Electric Power Research Institute, Palo Alto, CA (publication 1993).

11 Nalco Chemical Company, 1979, *The Nalco Water Handbook*, McGraw-Hill, New York.

12 Ostroff, A.G., 1979, *Introduction to Oilfield Water Technology*, National Association of Corrosion Engineers, Houston, TX.

13 Gary, J.F., Garey, J., Jorden, R., Aitken, A., Burton, D., and Gray, R., 1980, ed., *Condenser Biofouling Control*, Ann Arbor Science Publications, Ann Arbor, MI.

14 McCoy, J.W., 1980, *Microbiology of Cooling Water*, Chemical Publishing, NY.

15 ASM Metals Handbook, 1982, Vol. 5, *Surface Cleaning, Finishing, and Coating*, ASM International, Metals Park, OH.

References

1 ASM Metals Handbook, 1987, Vol. 13, *Corrosion*, ASM International, Metals Park, OH, p. 377.

2 ASM Metals Handbook, 1987, Vol. 13, *Corrosion*, ASM International, Metals Park, OH, p. 17.

3 Landrum, R.J., 1989, *Designing for Corrosion Control*, National Association of Corrosion Engineers, Houston, TX.

4 Licina, G.J., 1988, *Sourcebook for Microbiologically Influenced Corrosion in Nuclear Power Plants*, EPRI NP-5580, Electric Power Research Institute, Palo Alto, CA, p. 5-2.

5 ASM Metals Handbook, 1987, Vol. 5, *Surface Cleaning, Finishing, and Coating*, ASM International, Metals Park, OH, p. 83.

6 Licina, G.J., 1988, *Detection and Control of Microbiologically Influenced Corrosion*, EPRI NP-6815-D, Electric Power Research Institute, Palo Alto, CA, p. 5-7.

7 Lutey, R.W., 1989, *Microbiologically Influenced Corrosion in the Electric Power Generating Industry: A Training Program on Applied Technology*, Electric Power Research Institute, Palo Alto, CA.

8 ASM Metals Handbook, 1987, Vol. 5, *Surface Cleaning, Finishing, and Coating*, ASM International, Metals Park, OH, p. 65.

9 Casey, G.E., 1988, 'Introduction to chemical cleaning of service water systems', *EPRI Service Water Reliability Improvement Seminar*, Electric Power Research Institute, Palo Alto, CA.

10 NACE, 1984, *Corrosion Basics, An Introduction*, National Association of Corrosion Engineers, Houston, TX, p. 149.

11 Ostroff, A.G., 1979, *Introduction to Oilfield Water Technology*, National Association of Corrosion Engineers, Houston, TX, p. 156.

12 Dillon, C.P., 1986, *Corrosion Control in the Chemical Process Industries*, McGraw-Hill, New York, p. 160.

13 Tanis, J.N., 1987, *Procedures of Industrial Water Treatment*, Ltan, Ridgefield, CT, p. 1.

14 Puckorious, P., personal communication.

15 Betz, 1991, *Betz Handbook of Industrial Water Conditioning*, Betz Laboratories, Trevose, PA, p. 19.

16 Betz, 1991, *Betz Handbook of Industrial Water Conditioning*, Betz Laboratories, Trevose, PA, p. 23.

17 Betz, 1991, *Betz Handbook of Industrial Water Conditioning*, Betz Laboratories, Trevose, PA, p. 31.

18 Betz, 1991, *Betz Handbook of Industrial Water Conditioning*, Betz Laboratories, Trevose, PA, p. 38.

19 Betz, 1991, *Betz Handbook of Industrial Water Conditioning*, Betz Laboratories, Trevose, PA, p. 47.

20 NACE, 1984, *Corrosion Basics, An Introduction*, National Association of Corrosion Engineers, Houston, TX, p. 162.

21 Ostroff, A.G., 1979, *Introduction to Oilfield Water Technology*, National Association of Corrosion Engineers, Houston, TX, p. 336.

22 Betz, 1991, *Betz Handbook of Industrial Water Conditioning*, Betz Laboratories, Trevose, PA, p. 247.

23 NACE, 1984, *Corrosion Basics, An Introduction*, National Association of Corrosion Engineers, Houston, TX, p. 154.

24 Betz, 1991, *Betz Handbook of Industrial Water Conditioning*, Betz Laboratories, Trevose, PA, p. 275.

25 Betz, 1991, *Betz Handbook of Industrial Water Conditioning*, Betz Laboratories, Trevose, PA, p. 279.

26 Ostroff, A.G., 1979, *Introduction to Oilfield Water Technology*, National Association of Corrosion Engineers, Houston, TX, p. 270.

27 Ostroff, A.G., 1979, *Introduction to Oilfield Water Technology*, National Association of Corrosion Engineers, Houston, TX, p. 191.

28 ASM Metals Handbook, 1987, Vol. 13, *Corrosion*, ASM International, Metals Park, OH, p. 483.

29 Jordan, R.M., and Shearer, L.T., 1962, 'Aqualin biocide in injection waters', paper no. 280, presented at the research meeting, Tulsa, OK, Society of Petroleum Engineers.

30 Baumgartner, A.E., 1962, 'Microbiological corrosion – what causes it and how can it be controlled', *Journal of Petroleum Technology* **14**(10) 1074.

31 Sharpley, J.M., 1966, *Elementary Petroleum Microbiology*, Gulf Publishing, Houston, TX.

32 Ostroff, A.G., 1979, *Introduction to Oilfield Water Technology*, National Association of Corrosion Engineers, Houston, TX, p. 161.

33 Betz, 1991, *Betz Handbook of Industrial Water Conditioning*, Betz Laboratories, Trevose, PA, p. 200.

34 Betz, 1991, *Betz Handbook of Industrial Water Conditioning*, Betz Laboratories, Trevose, PA, p. 194.

35 McCoy, J.W., 1980, *Microbiology of Cooling Water*, Chemical Publishing, NY.

36 ASM Metals Handbook, 1987, Vol. 13, *Corrosion*, ASM International, Metals Park, OH, p. 493.

37 Betz, 1991, *Betz Handbook of Industrial Water Conditioning*, Betz Laboratories, Trevose, PA, p. 203.

38 Betz, 1991, *Betz Handbook of Industrial Water Conditioning*, Betz Laboratories, Trevose, PA, p. 206.

39 Betz, 1991, *Betz Handbook of Industrial Water Conditioning*, Betz Laboratories, Trevose, PA, p. 207.

40 Dillon, C.P., 1986, *Corrosion Control in the Chemical Process Industries*, McGraw-Hill, New York, p. 263.

41 NACE, 1984, *Corrosion Basics, An Introduction*, National Association of Corrosion Engineers, Houston, TX, p. 127.

42 NACE, 1973, *Corrosion Inhibitors*, National Association of Corrosion Engineers, Houston, TX, p. 11.

43 ASM Metals Handbook, 1987, Vol. 13, *Corrosion*, ASM International, Metals Park, OH, p. 494.

44 ASM Metals Handbook, 1987, Vol. 13, *Corrosion*, ASM International, Metals Park, OH, p. 482.

45 NACE, 1984, *Corrosion Basics, An Introduction*, National Association of Corrosion Engineers, Houston, TX, p. 131.

46 Betz, 1991, *Betz Handbook of Industrial Water Conditioning*, Betz Laboratories, Trevose, PA, p. 89.

47 NACE, 1984, *Corrosion Basics, An Introduction*, National Association of Corrosion Engineers, Houston, TX, p. 482.

48 Betz, 1991, *Betz Handbook of Industrial Water Conditioning*, Betz Laboratories, Trevose, PA, p. 185.

49 Licina, G.J., 1988, *Sourcebook for Microbiologically Influenced Corrosion in Nuclear Power Plants*, EPRI NP-5580, Electric Power Research Institute, Palo Alto, CA, p. 7-8.

50 Licina, G.J., 1988, *Detection and Control of Microbiologically Influenced Corrosion*, EPRI NP-6815-D, Electric Power Research Institute, Palo Alto, CA, p. 6-2.

51 Licina, G.J., 1988, *Detection and Control of Microbiologically Influenced Corrosion*, EPRI NP-6815-D, Electric Power Research Institute, Palo Alto, CA, p. 7-3.

52 Licina, G.J., 1988, *Detection and Control of Microbiologically Influenced Corrosion*, EPRI NP-6815-D, Electric Power Research Institute, Palo Alto, CA, p. 6-8.

53 Shin, S.W., and McEnerney, J.W., 1986, 'MIC in stainless steel pipe after hydrotesting', Workshop on *Microbe Induced Corrosion*, Electric Power Research Institute, Palo Alto, CA.

54 Puckorius, P.R., 1983, 'Massive condenser failure caused by sulfide producing bacteria', *Materials Performance* **22**(12) 19.

55 Puckorius, P.R., 1993, 'Microorganisms in cooling water systems – case histories: detrimental and beneficial effects', *Corrosion/93*, National Association of Corrosion Engineers, Houston, TX.

56 Licina, G.J., 1988, *Detection and Control of Microbiologically Influenced Corrosion*, EPRI NP-6815-D, Electric Power Research Institute, Palo Alto, CA, p. 6-9.

57 Borenstein, S.W., 1988, 'Microbiologically influenced corrosion failures of austenitic stainless steel welds', *Materials Performance* **27**(3) 62.

58 Garner, A., 1982, *Materials Performance* **20**(8) 9.

59 Kolts, J., and Sridhar, N., 1985, 'Temperature effects in localized corrosion', in *Corrosion of Nickel-Base Alloys*, ASM International, Metals Park, OH.

60 Licina, G.J., 1988, *Detection and Control of Microbiologically Influenced Corrosion*, EPRI NP-6815-D, Electric Power Research Institute, Palo Alto, CA, p. 6-3.

61 United States Nuclear Regulatory Commission IE Information Notice No. 85-24, March 26, 1985, 'Failure of protective coatings in pipes and heat exchangers'.

62 Licina, G.J., 1988, *Detection and Control of Microbiologically Influenced Corrosion*, EPRI NP-6815-D, Electric Power Research Institute, Palo Alto, CA, p. 6-4.

63 Tatnall, R.E., 1981, 'Case histories: bacteria induced corrosion', *Materials Performance* **20**(8) 32.

64 Kobrin, G., 1986, 'Reflections on microbiologically induced corrosion', in *Biologically Induced Corrosion*, Dexter, S.C., ed., National Association of Corrosion Engineers, Houston, TX.

65 Bibb, M., and Hartman, K.W., 1984, 'Bacterial corrosion', *Corrosion and Coatings South Africa*, October, 12.

66 King, R.A., Miller, J.D.A., and Scott, J.F.D., 1986, 'Subsea pipelines: internal and external biological corrosion', in *Biologically Induced Corrosion*, Dexter, S.C., ed., National Association of Corrosion Engineers, Houston, TX.

67 Gilbert, R.J., and Lovelace, D.W., 1975, *Microbial Aspects of the Deterioration of Metals*, Academic Press, London.

68 Mogil'nitskii, Zinevich, A.M., Borisov, B.I., Gracheva, T.B., Korobtsova, N.G., Chepizhenko, D.N., and Kerimov, S.I., 1980, 'Corrosion of polyvinyl chloride coated metal under the influence of biogenic hydrogen sulfide', translated from *Zaschita Metallov* **16**(2) 167.

69 Mussalli, Y.G., and Tsou, J., 1988, 'Nontoxic fouling control technologies: US and Japanese perspectives', EPRI Service Water Reliability Improvement Seminar, Electric Power Research Institute, Palo Alto, CA.

70 Sommerville, D.C., Borenstein, S.W., and Viswanath, B.P., 1988, 'Fouling control using non-toxic coatings', EPRI Service Water Reliability Improvement Seminar, Electric Power Research Institute, Palo Alto, CA.

71 EPRI, 'Nontoxic foul-release coatings', 1989, EPRI GS-6566, Electric Power Research Institute, Palo Alto, CA.

72 Miller, J.D.A., 1980, 'Principles of microbial corrosion', *British Corrosion Journal* **15**(2) 92.

73 Sharpley, J.M., 1973, 'Microbiological corrosion and its control', in *Corrosion Inhibitors*, Nathan, C.C., ed., National Association of Corrosion Engineers, Houston, TX.

74 Tiller, A.K., 1986, 'A review of the European research effort on microbial corrosion between 1950 and 1984', in *Biologically Induced Corrosion*, Dexter, S.C., ed., National Association of Corrosion Engineers, Houston, TX.

75 McDougal, J., 1966, 'Microbial corrosion of metals', *Anti-Corrosion*, August, 9.

76 Guthrie, P., 1987, *The Diversity of TVA*, Tennessee Valley Authority, Knoxville, TN.

Chapter 9

Testing

In this chapter, we will cover test methods related to MIC. As with any test, the purpose of testing must be clearly defined. Since corrosion behaviour is usually site-specific, resulting from the environment of exposure as well as the properties of the material, no all-purpose corrosion test exists – nor should it, despite numerous misunderstandings to the contrary. Many failures have occurred because engineers used published data of 0.38 mm (15 mils) per year general corrosion rate, for example, in a specific environment for a specific material, and the actual service conditions led to failure by pitting corrosion owing to pitting rates of greater than 1.27 mm (50 mils) per year. Literature commonly gives general corrosion rates for many situations. It must be emphasized that the response of a material to environments where pitting or crevice corrosion conditions exists, and what the pitting rate will be, is *not* reflected in the general corrosion rate data.

Similarly, there is no standard test for determining the likelihood for MIC to occur.[1] The presence or absence of microorganisms means little in terms of MIC. It may be looked upon as a trend. The microorganisms may or may not influence the corrosion process. For example, MIC damage may have occurred during an outage, often years before a failure is observed. When the failure is finally examined, the organisms are no longer viable. Before requesting testing, ask what the course of action will be if the test is positive and if the test is negative. As a general rule, if the answer is the same, do not do the test.

Pitting, for example, may have progressed at a rapid rate for the first year and then slowed to a minor rate for subsequent years, until failure occurred. The rate (pit depth divided by the number of years in service) is *not* necessarily linear. Pitting rates may be erratic and initiation times difficult to pinpoint.

One good source of information about MIC testing is the *Proceedings of the International Symposium on Microbiologically Influenced Corrosion (MIC) Testing* sponsored by ASTM Committee G-1 on Corrosion of Metals held in Miami, Florida, in November 1992.

Table 9.1. Typical test objectives[2]

- Determining the best material of construction for a specific environment
- Predicting the probable service life
- Evaluating new alloys or new processes
- Evaluating corrosion-resistant materials in different process conditions, e.g. pH
- Evaluating corrosion-resistant materials in different control conditions, e.g. inhibitors
- Investigating corrosion mechanisms
- Estimating corrosion rate

9.1   Test objectives

Corrosion tests often reduce costs. Having the right material in the right environment saves money, avoids lost operating time and decreases the need for repair and replacement. Determine clear objectives and the criteria for interpreting test results.[2] Typical test objectives are given in Table 9.1. Statistics are often used with corrosion test results and in a well-designed test, information will be statistically valid. You will also need specific acceptance criteria, as well as a clearly defined test method.

There are two categories of tests, laboratory and field tests. The specific situation may determine which is appropriate, for example, for a quality assurance concern, you would need a laboratory.

A sizable amount of research in MIC is fundamental, focusing on how or why a failure or, more specifically, a corrosion mechanism occurs. Standardized corrosion test methods are often specialized, while standardized tests relating to MIC are typically unavailable. Progress has been made in standardizing some test practices or methods, for example, in electrochemical testing.[3–6]

9.1.1   Material factors

There are many factors to consider in determining the corrosion behaviour of a material. Because materials are used in so many conditions, you need to consider a complex set of parameters to predict performance. The materials themselves are often modified during service, such as by maintenance and repair. The service conditions often change as well, such as seasonal changes of water and salinity content, or temperature variations.

Some factors relate to the advantages and limitations of metallurgical and corrosion characteristics of metals and alloys.[7] Typical material factors include composition, homogeneity, stress and thermal history.

The chemical composition of a metal or alloy influences its corrosion behaviour. The specific composition, rather than the commercially available range, may be extremely important. For laboratory testing, it is common to obtain a chemical analysis of the material being tested.

Generic information, for example, such as the material's grade is often not sufficient for many situations.

Materials are often erroneously considered homogeneous. For example, a Type 304L stainless steel plate would have a specific chemistry, and a heat number from the manufacturer might be available. But consider two pieces of plate welded together: each may have a different chemistry, the weld metal and the associated weld zone will themselves have different chemistries, and the heat of welding would have produced variations at different locations. For MIC, the concept of homogeneity simply does not apply.

To a microorganism, surface features such as oxide films or scratches in a metal surface, for example, may be orders of magnitude larger than they are, and may strongly influence adhesion. These surface features may dominate in the process of microbial settlement of the surface. After a film or colony of microbes begins to form, such features may dramatically influence corrosion, as discussed previously.

Stress, such as residual stress from cold working, can affect general corrosion.[8] Thermal history also influences corrosion. Materials that have been annealed, stabilized, sensitized, stress-relieved or welded, for example, will behave differently in specific environments.[8] These effects should be considered during corrosion testing, especially MIC testing.

9.2   Test results

Corrosion testing methods are often criticized because of test interpretations. Test results may indicate a favourable situation for a specific set of parameters. Engineers apply that information in the design or prediction of service life for a component or system.[9] Unfortunately, the actual corrosion rates in service may be higher than the predicted rates, with disastrous results.

Use corrosion rates with caution.[10] Short-term tests may provide invalid information if corrosion does not occur at a linear rate. Generally, this is the exception rather than the rule, and short-term tests do successfully predict long-term exposures. As discussed in Chapter 4, corrosion is also dependent on surface area, local composition, heat treatment variations, crevices and many other factors.

The corrosion rate may vary with the effect of film formation on the metal surface. The film forms a barrier of corrosion products that may protect the metal surface from a corrosive environment, and therefore reduces the corrosion rate over time. A short-term test may miss the effects of the film forming and thereby miss a decrease in the corrosion rate.

9.3   Laboratory testing

Laboratory testing is almost always performed under controlled conditions, using multiple specimens and carefully monitored parameters. These tests are generally deemed reproducible.

Laboratory tests are often made with small specimens, and are intended to evaluate the corrosion-response characteristics of the metal.[11] Tests may be used, for example, to evaluate how several materials will perform under the same conditions before a material is selected for construction. Many standardized tests are described in the annual book of ASTM Standards, *Metal, Corrosion, and Wear*, Vol. 03.02.

9.3.1   Electrochemical methods

Electrochemical methods work well with testing corrosion, since corrosion is an electrochemical process. Not only do chemical reactions occur, but a transfer of electrons also takes place. Several textbooks cover electrochemical testing methods in corrosion.[12–15] (See Section 4.4.2 on the fundamentals of electrochemistry, and for polarization diagrams.) In addition, Mansfeld and Little reviewed electrochemical techniques for studying MIC; both papers are excellent sources.[3, 16] Note that Little has concluded that details in testing have an extensive impact on test results.[17]

Polarization is the change in the open circuit electrode potential as the result of the passage of current. The results are often plotted on a polarization curve, showing the current density versus electrode potential with a given electrode and electrolyte, as shown in Fig. 9.1.[18] The x-axis in the polarization diagram is log I and the y-axis is E (potential). The value log I is current density, or current per unit area. The value E_{corr} (also called the open circuit potential, OCP) is the potential when no current is flowing through the cell (made up of an anode, a cathode, an electrolyte and a circuit). The point E_{corr} is where the anode potential and the cathode potential intersect. I_{corr} is the current density at E_{corr} or at the free corrosion potential.

Figure 9.2 illustrates that, as the corrosion reaction progresses, the two reactions, anodic and cathodic, drift towards each other until they intersect. This intersection is E_{corr} and I_{corr}. The rate of corrosion is directly proportional to current flow, and $E_{cell} = E_{cathode} - E_{anode}$.

Three tests are commonly used for the study of localized corrosion: polarization resistance, potentiostatic and potentiodynamic polarization.

- Polarization resistance testing is based on the slope (dE/dI) of the corrosion potential on a potential (E) versus current density (I) curve. The term polarization resistance describes both the slope and the test method.

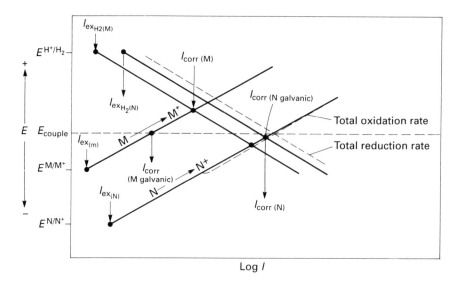

9.1 Potential–current relationships for the case of a galvanic couple between two corroding metals. M, more noble metal, N, less noble metal (source: Fontana, M., and Greene, N.D., 1978, *Corrosion Engineering*, © 1978 McGraw-Hill, New York, reproduced with permission).

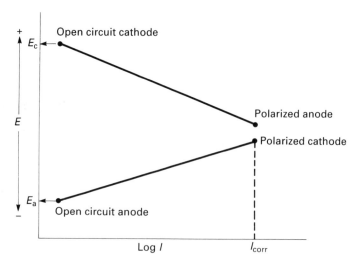

9.2 Polarization in a corrosion cell. $E_{cell} = E_{cathode} - E_{anode}$.

- Potentiostatic polarization is measured by an electrochemical test in which the electrode potential is held constant.
- Potentiodynamic polarization is measured by an electrochemical test in which the potential of an electrode is varied in a continuous manner at a preset rate.

Polarization techniques, such as potentiostatic and potentiodynamic polarization, are used to determine basic information about the following:[19]

- Passivity.
- Susceptibility to pitting.
- Susceptibility to crevice corrosion.
- The effects of alloying.
- The use of inhibitors on the kinetics of a reaction.

Corrosion potential

Measuring the corrosion potential E_{corr} is easy, but gives little information about the corrosion mechanism. Use of a reference electrode and a high impedance voltmeter to measure the corrosion potential is suitable for some corrosion cells (Fig. 9.3). When the topic is MIC, the system is more complex, owing to the formation of a biofilm.[3] The interpretation is difficult. Mansfeld and Little advise determining not only E_{corr}, but also simultaneously the polarization resistance, R_p.[3] The change in both parameters indicates changes in the rates of the anodic and/or cathodic reactions, which determine the corrosion rate.

For example, measurements of E_{corr} of steel in the presence of SRBs lack specific information, which makes interpretation difficult. The observed changes led to a variety of theories. The changes to the negative direction of E_{corr} resulted from a reduction of the cathodic reaction rate and were observed as the result of an increase in the anodic reaction rate.[20, 21] Both theories lead to the same result, but without additional data, no valid conclusions can be drawn. Mansfeld and Little agree that without information about the exact mechanisms of MIC for a given system, one cannot use E_{corr} data for monitoring purposes, such as the detection of an increase in uniform corrosion rates, or the initiation of localized corrosion due to bacteria.

The case of stainless steels in seawater is another example of possible over-interpretation of reports on the time-dependence of E_{corr}. The reports show an ennoblement of E_{corr} during exposure.[3, 22-29] However, from the data, one cannot tell if the increase in E_{corr} is from thermodynamic effects, kinetic effects or both. The oxygen reduction can increase because of a local increase in oxygen concentration, or a decrease in pH. As Mansfeld and Little note, the situation becomes complex, since naturally occurring microorganisms within a marine biofilm can either increase or decrease the local oxygen concentration and pH.[3] Little et al.

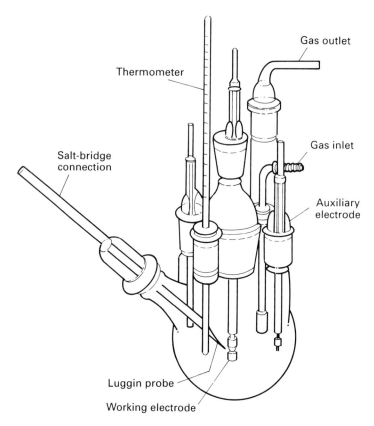

9.3 Typical electrochemical polarization cell illustrating locations for working and auxiliary electrodes, and associated cell components.

have reported some of the first measurements of interfacial chemistry, and have demonstrated its impact on the development of E_{corr} with exposure time.[17,30] They showed that increasing the exchange current density for the oxygen reduction reaction would tend to increase E_{corr}. Since E_{corr} is a mixed potential, its value may change due to the passive current density of the stainless steel. At constant kinetics of the oxygen reduction reaction, E_{corr} would become more noble if the passive current density decreases with time. Clearly, this shows how simply measuring E_{corr} is not enough to enable actual corrosion mechanisms to be interpreted.

Redox potential

The reduction–oxidation (redox), or solution, potential indicates the oxidation power of an electrolyte. It is usually measured with an inert electrode such as platinum. If suitable calibration is provided, redox could be used to indicate the corrosivity of an electrolyte. Mansfeld and Little note that although it has been used to monitor changes in a solution of a given corrosivity as a result of bacterial metabolism, redox can produce varying degrees of corrosion for different metals such as mild steel, stainless steel or aluminium exposed to the same solution. Redox may be useful in combination with measurements of E_{corr} and R_p for monitoring MIC.[3]

Polarization resistance technique

Polarization resistance is commonly used to measure uniform corrosion. When the anodic and cathodic polarization is within 10 mV of the corrosion potential, Bockris and others showed that the applied current density is approximately linear with potential.[13,31] The slope of the linear curve, E-log I, is the polarization resistance R_p (in units of resistance, ohms). Stern and Geary derived an expression showing that the corrosion rate is inversely proportional to R_p at potentials close to E_{corr}.[32,33]

Mansfeld gives a complete review of using this technique for measuring corrosion current, which allows for continuous monitoring of the instantaneous corrosion rate of a metal in a specific environment. Mansfeld and Little discuss the technique's suitability for detecting changes in corrosion rates due to the presence of bacteria, inhibitors, sunlight, biocides and so forth. The problem again lies with interpretation of the data.

The polarization resistance technique is simplified as the linear polarization resistance technique (Fig. 9.4), in which the relationship between E

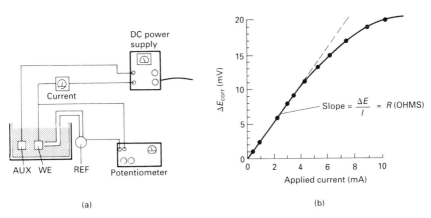

9.4   (a) Circuit for conducting linear polarization measurements; (b) typical data obtained (source: NACE, 1984, *Corrosion Basics: An Introduction*, NACE, Houston, TX, reproduced with permission).

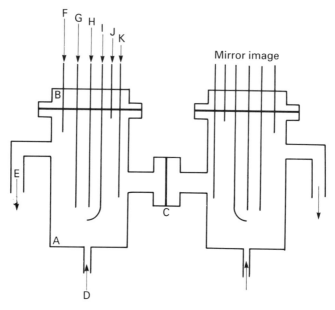

9.5 Two compartment corrosion cell (dual cell) used to evaluate microbial corrosion: A half cell; B cell cover; C membrane; D medium inlet; E medium outlet; F utility point; G Pt electrode; H metal specimen; I luggin capillary; J gas outlet; K gas inlet (source: Little, B.J., Wagner, P.A., Gerkachov, S.M., Walch, M., and Mitchell, R., 1986, *Corrosion* 42(9) 533, reproduced with permission).

and *i* is assumed to be linear (rather than merely close to linear).[33,34] Problems occur with use in MIC testing because the interface behaves as a resistance with the magnitude inversely proportional to the corrosion current.[32] Little *et al.* discuss the interface's actual behaviour as a parallel resistance–capacitance combination because of chemical reactions, solvation and the absorption of reaction intermediates, which results in a very complicated system to model and interpret.

Dual-cell technique

Little *et al.* developed the dual-cell technique to study MIC.[35,36] This technique allows for continuous monitoring of the changes in the corrosion rate of a metal due to the presence of a biofilm. The technique uses two electrochemical cells (Fig. 9.5) separated by a semipermeable membrane. The cells are identical; one cell is for the addition of bacteria and the other the control. The measurement of galvanic current provides a continuous record of the time-dependence of the corrosion process. It does not allow calculation of the corrosion rate. Little *et al.* have used the dual-cell technique to simulate the oxygen-deficient or oxygen-free conditions under biofilms as well as other applications.[36–39]

Electrochemical impedance spectroscopy

The electrochemical impedance spectroscopy (EIS) technique, sometimes referred to as AC impedance, uses a small amplitude AC signal applied to the test electrode; the researcher determines a response as a function of the frequency of the signal.[31] The background of EIS and its applications to electrochemistry may be found in the proceedings of the First International Symposium on EIS.[40] Mansfeld, Silverman and MacDonald have all published basic concepts regarding the EIS technique.[34, 41–43]

EIS provides a different type of information from that given by the other techniques discussed above. Using EIS, a technician analyses the corrosion system at a fixed potential (or current). The properties of the system at this potential can be determined through the analysis of the frequency dependence of the impedance.

EIS is used for studying MIC. Mansfeld and Little used the technique to examine stainless steel and titanium grade 2 during exposure to natural seawater.[44] They found that the small decrease of the capacitance over time could be due to the formation of the biofilm, which examination, using the SEM, showed to be of a patchy nature. Mansfeld and Shih have demonstrated the usefulness of using EIS for detecting the initiation and growth of pits in aluminium-based metal matrix composites.[45]

In practice, there are problems in using EIS to study MIC.[33] The technique is not necessarily easy to apply, and the data are difficult to interpret.[19] The charge transfer resistance is used to determine corrosion rate. Little *et al.* note that other resistances, such as solution resistance, film resistance and capacitance, are related to the reaction rates between the electrode and the electrolyte.[32] When the corrosion product becomes bulky, as is frequently the case with MIC, the impedance increases and the accuracy of the calculations decreases.

Electrochemical noise analysis

The electrochemical noise technique measures fluctuations of the potential, often E_{corr}, or the current as a function of time. This technique is described by Mansfeld and Little; they suggest that it is more suited for monitoring purposes than mechanistic studies at the present stage of development.[3]

Researchers at the University of Manchester Institute of Science and Technology (UMIST) have used this technique to study MIC.[46] They studied the effect of SRBs and other organisms on reinforcing steel and concrete and concluded that the mechanism of corrosion of reinforcement steel in concrete by SRBs was a stepwise process, in which the passive oxide film was first destroyed, followed by a continuous rupture and the formation of an initially protective but ultimately destructive film of iron sulphide. This film contributed to pitting corrosion.

Large signal polarization techniques

The large signal polarization differs from those discussed above in that this technique uses potential scans that range from several hundred millivolts to several volts.[3] Mansfeld and Little discuss several specific cases of investigators using this technique.[3] There seems to be no agreement among authors studying MIC regarding this technique's value in investigating the corrosion mechanisms, which is probably due partly to the complexity and partly to the variety of conditions and materials examined. The authors do seem to agree that the observed increase of corrosion rates is due to the influence of microorganisms on the rates of cathodic and anodic reactions involved with the corrosion process.

9.3.2   Other methods

Other methods for laboratory testing include immersion tests, salt spray testing and miscellaneous testing, such as simulated atmosphere testing. These tests are described in the chapter on laboratory testing in the ASM Metals Handbook, 1987, Vol. 13, *Corrosion*. None of these tests lends itself very well to the study of MIC.

9.4   Field testing

The field test is useful for determining suitability for service when duplicating operating conditions in the laboratory would be difficult, whether due to pressures, temperatures, flows, contaminants or other factors. The disadvantages are that specific conditions such as weldments, crevices, stresses and so forth may affect corrosion mechanisms and not be reproducible in field testing.

9.4.1   Corrosion testing

A variety of techniques for corrosion testing are available for on-site monitoring.[47–50] Corrosion testing is often done using weight loss coupons. Many coupons of various materials can be exposed simultaneously, and the materials can then be ranked as to their performance after exposure to an actual process.[47] There are four primary limitations to coupon testing:[47]

- Coupon testing cannot be used to detect rapid changes in corrosivity.
- Corrosion rates on coupons do not duplicate corrosion rates on equipment.
- The rate calculated from the coupon cannot be taken directly as the corrosion rate of the equipment.
- Coupons do not detect some forms of corrosion.

In terms of MIC testing, sidestream loop tests are sometimes done, usually to optimize operating parameters or to evaluate various control treatments simultaneously.[51] Characklis notes that although simulating operating equipment in every detail with a sidestream test loop is virtually impossible, important parameters (i.e. water flow velocity, tube material and heat flux) can be matched to ensure that a realistic, effective treatment regime can be developed.

Videla has developed an experimental sampling device that can be inserted directly into a flow stream or on a side stream conventional corrosion rack.[52] The sampler, called Renaprobe™ (REusable Nonconventional Appliance), consists of a Teflon™ rod mounted with metal coupons, each of $0.5\,cm^2$ $(0.075\,in^2)$ surface area.

Traditional corrosion monitoring probes (e.g. Corrator™ and Corrosometer™, tradenames of Rohrback Cosasco) do not seem to be suitable for directly determining information about MIC, most likely because the size of the coupon is too small for MIC to initiate.[53]

9.4.2 Electrochemical testing

New designs for probes and monitors are being developed to allow field testing and on-line monitoring for biofilm activity and to provide a means for MIC control. Use of monitors should also improve methods for injecting biocides to avoid over-treatment, to reduce costs associated with biocides and to reduce the amount of biocides in the environment.

The BI·GEORGE™ electrochemical biofilm monitor, which permits on-line monitoring of biofilm activity, is currently commercially available (patent pending, EPRI 1991; tradename of Structural Integrity Associates).[6,54,55] The probe, designed specifically for monitoring biofilm activity in industrial applications, has been tested in freshwater environments in the laboratory and in the field. The probe monitors changes in electrochemical reactions produced by biofilms on stainless steel electrodes. The probe (Fig. 9.6) consists of:

- A pair of identical electrodes, with each electrode comprising a series of identical stainless steel discs.
- A threaded body for attachment to the system to be monitored.
- A simple control and data acquisition system.

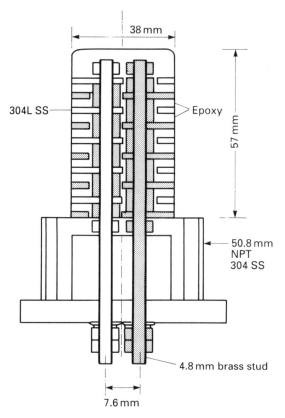

9.6 Electrochemical biofilm monitoring probe (from Licina, G.J., and Nekoksa, G., 1993, *Corrosion/93*, paper no. 403, NACE, Houston, TX, reproduced with permission).

One electrode is cathodically polarized for a short time (30 minutes to one hour) each day. The influence of biofilms on half-reactions is detected by monitoring the current required to achieve the pre-set potential over a period of days or weeks. The intermittent polarization produces conditions that result in differences in both the types and numbers of microorganisms present on each electrode. The polarization simulates electrochemical conditions similar to those resulting from local anodic sites. The mild polarization also encourages microbial colonization, similar to that observed on cathodically protected structures in microbially active environments.[56, 57]

Suggested reading

1 NACE, 1984, *Corrosion Basics, An Introduction*, National Association of Corrosion Engineers, Houston, TX.

2 Characklis, W.G., and Marshall, K.C., 1990, *Biofilms*, John Wiley & Sons, New York.

3 Dillon, C.P., 1986, *Corrosion Control in the Chemical Process Industries*, McGraw-Hill, New York.

4 ASM Metals Handbook, 1987, Vol. 13, *Corrosion*, ASM International, Metals Park, OH.

5 Ostroff, A.G., 1979, *Introduction to Oilfield Water Technology*, National Association of Corrosion Engineers, Houston, TX.

6 Ailor, W.H., 1971, *Handbook on Corrosion Testing and Evaluation*, John Wiley & Sons, New York.

7 Haynes, G.S., and Baboian, R., ed., 1985, *Laboratory Corrosion Tests and Standards*, STP 866, American Society for Testing and Materials.

8 Dexter, S.C., ed., 1986, *Biologically Influenced Corrosion*, NACE Reference Book No. 8, National Association of Corrosion Engineers, Houston, TX.

9 Licina, G.J., 1988, *Sourcebook for Microbiologically Influenced Corrosion in Nuclear Power Plants*, EPRI NP-5580, Electric Power Research Institute, Palo Alto, CA.

10 Licina, G.J., 1988, *Detection and Control of Microbiologically Influenced Corrosion*, EPRI NP-6815-D, Electric Power Research Institute, Palo Alto, CA.

References

1 ASM Metals Handbook, 1987, Vol. 13, *Corrosion*, ASM International, Metals Park, OH, p. 314.

2 ASM Metals Handbook, 1987, Vol. 13, *Corrosion*, ASM International, Metals Park, OH, p. 193.

3 Mansfeld F., and Little, B.J., 1990, 'The application of electrochemical techniques for the study of MIC – a critical review', *Corrosion/90*, paper no. 108, National Association of Corrosion Engineers, Houston, TX.

4 Dexter, S.C., Siebert, O.W., Duquette, D.J., and Videla, H., 1989, 'Use and limitations of electrochemical techniques for investigating microbiological corrosion', *Corrosion/89*, paper no. 616, National Association of Corrosion Engineers, Houston, TX.

5 Mollica, A., Trevis, A., Traverso, E., Ventura, G., DeCarolis G., and Dellepianne, R., 1989, *Corrosion* **45** 786.

6 Licina, G.J., and Nekoksa, G., 1993, 'On-line monitoring of microbiologically influenced corrosion in power plant environments', *Corrosion/93*, paper no. 403, National Association of Corrosion Engineers, Houston, TX.

7 Dillon, C.P., 1986, *Corrosion Control in the Chemical Process Industries*, McGraw-Hill, New York, p. 73.

8 Dillon, C.P., 1986, *Corrosion Control in the Chemical Process Industries*, McGraw-Hill, New York, p. 74.

9 ASM Metals Handbook, 1987, Vol. 13, *Corrosion*, ASM International, Metals Park, OH, p. 95.

10 ASM Metals Handbook, 1987, Vol. 13, *Corrosion*, ASM International, Metals Park, OH, p. 195.

11 NACE, 1984, *Corrosion Basics, An Introduction*, Houston, TX, p. 309.

12 ASM Metals Handbook, 1987, Vol. 13, *Corrosion*, ASM International, Metals Park, OH, p. 228.

13 Bockris, J.O'M., 1954, *Modern Aspects of Electrochemistry*, Butterworths, London.

14 Fontana, M.G., and Greene, N.D., 1978, *Corrosion Engineering*, McGraw-Hill, New York.

15 Uhlig, H.H., and Revie, R.W., 1985, *Corrosion and Corrosion Control*, John Wiley & Sons, New York.

16 Mansfeld, F., and Little, B.J., 1991, *Corrosion Science*, **32** 247.

17 Little, B.J., Ray, R., Wagner, P., Lewandowski, Z., Lee, W.C., Characklis, W.G., and Mansfeld, F., 1990, 'Electrochemical behavior of stainless steels in natural seawater', *Corrosion/90*, paper no. 150, National Association of Corrosion Engineers, Houston, TX.

18 ASM Metals Handbook, 1987, Vol. 13, *Corrosion*, ASM International, Metals Park, OH, p. 217.

19 Mansfeld, F., 1988, *Corrosion* **44**(12) 856.

20 Hadley, R.F., 1943, 'The influence of *Sporovibrio desulfuricans* of the current and potential behavior of corroding iron', National Bureau of Standards Conference.

21 Wanklin, J.N., and Spriut, C.I.P., 1952, *Nature* **169** 928.

22 Mollica, A., Trevis, A., Traverso, E., Venture, G., Scotto, V., Alabisio, G., Marcenaro, G., Montini, U., deCarolis, G., and Dellepiane, R., 1984, 'Interaction between biofouling and oxygen reduction rate on stainless steel in seawater', *Proc. 6th Int. Cong. Marine Corrosion and Fouling*, Athens, Greece, p. 269.

23 Scotto, V., DeCintio, R., and Marcenaro, G., 1985, 'The influence of marine aerobic microbial film on stainless steel corrosion behaviour', *Corrosion Science* **25** 185.

24 Johnsen, R., and Bardel, E., 1985, 'Cathodic properties of different stainless steels in natural seawater', *Corrosion* **41** 296.

25 Johnsen, R., and Bardel, E., 1986, 'The effect of a microbiological slime layer on stainless steel in natural seawater', *Corrosion/86*, paper no. 227, National Association of Corrosion Engineers, Houston, TX.

26 Videla, H.A., deMele, M.F.L., and Brankevich, G., 1987, 'Microfouling of several metal surfaces in polluted seawater and its relation with corrosion', *Corrosion/87*, paper no. 365, National Association of Corrosion Engineers, Houston, TX.

27 Mollica, A., Trevis, A., Traverso, E., Ventura, G., DeCarolis, G., and Dellapiane, R., 1989, 'Cathodic performance of stainless steels in natural seawater as a function of microorganism settlement and temperature', *Corrosion* **45** 48.

28 Mollica, A., Trevis, A., Traverso, E., and Scotto, V., 1988, 'Cathodic behaviour of nickel and titanium in natural seawater', *Int. Biodeterior.* **24** 221.

29 Dexter, S.C., and Gao, G.Y., 1988, 'Effect of seawater biofilms on corrosion

potential and oxygen reduction of stainless steel', *Corrosion* **44** 717.

30 Lewandowski, Z., Lee, W.C., Characklis, W.G., and Little, B., 1989, 'Dissolved oxygen and pH microelectrode measurements at water immersed metal surfaces', *Corrosion/89*, paper no. 93, National Association of Corrosion Engineers, Houston, TX.

31 ASM Metals Handbook, 1987, Vol. 13, *Corrosion*, ASM International, Metals Park, OH.

32 Little, B.J., Wagner, P.A., Characklis, W.G., and Lee, W., 1990, 'Microbial corrosion', in *Biofilms*, Characklis, W.G., and Marshall, K.C., ed., John Wiley & Sons, New York.

33 Stern, M., and Geary, A.L., 1957, 'Electrochemical polarization', *J. Electrochemic. Soc.* **104** 33.

34 Mansfeld, F., 1981, *Corrosion* **37** 310.

35 Gerchakov, S.M., Little, B.J., and Wagner, P., 1986, 'Probing microbiologically induced corrosion', *Corrosion* **42** 689.

36 Little, B.J., Wagner, P., Gerchakov, S.M., Walch, M., and Mitchell, R., 1986, 'The involvement of a thermophilic bacterium in corrosion processes', *Corrosion* **42** 533.

37 Wagner, P., and Little, B.J., 1986, 'Applications of a technique for the investigation of microbiologically induced corrosion', *Corrosion/86*, paper no. 122, National Association of Corrosion Engineers, Houston, TX.

38 Little, B.J., and Wagner, P., 1986, 'An electrochemical evaluation of microbiologically induced corrosion by two iron-oxidizing bacteria', *Corrosion/86*, paper no. 122, National Association of Corrosion Engineers, Houston, TX.

39 Little, B.J., Wagner, P., and Gerchakov, S.M., 1986, 'A quantitative investigation of mechanisms for microbial corrosion', in *Biologically Induced Corrosion*, Dexter, S.C., ed., NACE Reference Book No. 8, National Association of Corrosion Engineers, Houston, TX.

40 1989, *Proc. First International Symposium on EIS*, Bombannes, France.

41 Mansfeld, F., Kendig, M.W., and Tsai, S., 1982, *Corrosion* **38** 570.

42 Silverman, D.C., 1986, 'Primer on the AC impedance technique', in *The Electrochemical Techniques for Corrosion Engineering*, Baboian, R., ed., National Association of Corrosion Engineers, Houston, TX.

43 MacDonald, D.D., 1978, *Impedance Spectroscopy*, John Wiley & Sons, New York.

44 Mansfeld, F., Tsai, C.H., Shih, H., Little, B., Ray, R., and Wagner, P., 1990, 'Results of exposure of stainless steels and titanium to natural seawater', *Corrosion/90*, paper no. 109, National Association of Corrosion Engineers, Houston, TX.

45 Mansfeld, F., and Shih, H., 1988, *J. Electrochem. Soc.* **135** 1171.

46 Moosavi, A.N., Dawson, J.L., and King, R.A., 1986, 'The effect of sulfate-reducing bacteria on the corrosion of reinforced concrete', in *Biologically Influenced Corrosion*, Dexter, S.C., ed., NACE Reference Book No. 8, National Association of Corrosion Engineers, Houston, TX.

47 ASM Metals Handbook, 1987, Vol. 13, *Corrosion*, ASM International, Metals Park, OH, p. 197.

48 Dillon, C.P., Krishner, A.S., and Wissenberg, H., 1971, 'Plant corrosion tests', in *Handbook on Corrosion Testing and Evaluation*, Ailor, W.H., ed.,

John Wiley & Sons, New York.

49 Moran, G.C., and Labine, P., ed., 1986, *Corrosion Monitoring in Industrial Plants Using Nondestructive Testing and Electrochemical Methods*, STP 908, American Society for Testing and Materials.

50 NACE, 1976, *Laboratory Corrosion Testing of Metals for the Process Industries*, NACE TM-01-69. National Association of Corrosion Engineers, Houston, TX.

51 Characklis, W.G., Little, B.J., Stoodley, P., and McCaughey, M.S., 1991, 'Microbial fouling and corrosion in nuclear power plant service water systems', *Corrosion/91*, paper no. 281, National Association of Engineers, Houston, TX.

52 Videla, H.A., Silva, R.A., Canales, C.G., and Wilkes, J.F., 1992, 'Monitoring biocorrosion and biofilms in industrial waters: a practical approach', in *Proceedings of the International Symposium on Microbiologically Influenced Corrosion (MIC) Testing*, sponsored by ASTM Committee G-1 on Corrosion of Metals held in Miami, Florida, in November 1992.

53 George Licina, personal communication.

54 Licina, G.J., Nekoksa, G., and Howard, R.L., 1992, 'An electrochemical method for on-line monitoring of biofilm activity in cooling water using the BI·GEORGE™ probe', in *Proceedings of the International Symposium on Microbiologically Influenced Corrosion (MIC) Testing*, sponsored by ASTM Committee G-1 on Corrosion of Metals held in Miami, Florida, in November 1992.

55 Licina, G.J., Nekoksa, G., and Howard, R.L., 1992, 'An electrochemical method for on-line monitoring of biofilm activity in cooling water', *Corrosion/92*, paper no. 177, National Association of Corrosion Engineers, Houston, TX.

56 Nekoksa, G., and Gutherman, B., 1992, *'Cathodic Protection Criteria for Controlling Microbially Influenced Corrosion in Power Plants*, EPRI NP-7312, Electric Power Research Institute, Palo Alto, CA.

57 Guezennec, J., Dowling, N., Conte, M., Antoine, E., and Fiksdal, L., 'Cathodic protection in marine sediments and the aerated seawater column', *'Microbially Influenced Corrosion and Biodeteriation*, Dowling, N.J.E., Mittlemen, M.W., and Danko, J.C., ed., National Association of Corrosion Engineers, Houston, TX.

Index

AC impedance technique, 275
Acid cleaning, 243
Acoustic emission testing, 190
Adenosine triphosphate (ATP) assay
  method, 208
Aeration, 248
Aerobic, 9
Aerobic corrosion, 31
Algae, 10, 12
Alloy steels
  effect of alloying elements, 60
  defined, 51
Aluminum and aluminum alloys, 103
  anodes, 230
Amines, as corrosion inhibitors, 252
Anaerobic, 9
Anaerobic corrosion, 31
Anions, 14
Annealing, 53
Anode
  comparison of sacrificial, 230
  galvanic, defined, 230
  defined, 114
Anodic polarization, 39, 119
Anodic reaction, 39, 43, 132, 152
Antibody tests, 208
APS reductase method, 208
Asiatic clams, 253
Austenite, 51
Austenitizing, 53
Autotroph, 9

Bacteria (*see also* microorganisms),
  Aerobacillus, 202
  Aerobacter, 33
  Bacillus, 33
  Beggiatoa, 30
  Clonothrix, 27
  Clostridium, 202
  Crenothrix, 27, 42, 202
  Desulfobacter, 23, 42
  Desulformaculum, 23
  Desulfovibrio, 23, 42

Escherichia, 33
Escherichia coli, 12, 16
Flavobacterium, 33, 202
Gallionella, 27, 31, 42, 162, 203
Legionella, 214
Leptothrix, 27, 42, 162
Pseudomonas, 28, 33, 42
Salmonella, 202
Shiggela, 202
Siderscapa, 203
Sphaerotilus, 27, 42, 202
Streptococcus, 202
Sulfolobus, 30
Thiobacillus, 30, 171, 174
Thiodendron, 30
T. ferroxidans, 31
T. thiooxidans, 31, 33, 43
  acid-producing, 31
  slime-formers, 33
  iron bacteria, 27, 29, 170
  iron-oxidizing bacteria, 27
  manganese bacteria, 27
  metabolic processes, 33
  sulphate-reducing, *see* sulphate-reducing
    bacteria
  sulphur-oxidizing bacteria, 30, 174
Bacteriocides, *see* biocides
Bainite, 51
Biochemical oxygen demand (BOD), 16
Biocides (*see also* corrosion inhibitors),
  classifications, 251
  defined, 11, 251
  oxygenating, 251
  used in cooling water systems, 252
Biocorrosion, 4
Biodegradation, 5
Biodeterioration, 5
Biofilm (*see also* slime),
  and chemical influences, 34
  as crevices, 31, 35, 146
  defined, 10
  developmental stages, 1, 36
  effect of flow velocity, 36